Electrochemistry and Electrochemical Engineering

Electrochemistry and Electrochemical Engineering

Edited by
Lenny Hart

Larsen & Keller
www.larsen-keller.com

Electrochemistry and Electrochemical Engineering
Edited by Lenny Hart
ISBN: 978-1-63549-099-2 (Hardback)

Larsen & Keller

Published by Larsen and Keller Education,
5 Penn Plaza,
19th Floor,
New York, NY 10001, USA

Cataloging-in-Publication Data

Electrochemistry and electrochemical engineering / edited by Lenny Hart.
p. cm.
Includes bibliographical references and index.
ISBN 978-1-63549-099-2
1. Electrochemistry. 2. Electrochemical analysis. 3. Electrochemistry, Industrial.
I. Hart, Lenny.
QD553 .E44 2017
541.37--dc23

Printed and bound in China.

For more information regarding Larsen and Keller Education and its products, please visit the publisher's website www.larsen-keller.com

Table of Contents

Preface

This book outlines the processes and applications of electrochemistry in detail. It includes detailed explanation of the various concepts and uses of the subject. Electrochemistry, as a branch of physical sciences, examines the relation between electricity and the chemical changes caused by it. Electrochemical engineering deals with the refining of metals as well as processes like electroplating. Such selected concepts that redefine this subject area have been presented in the text. While understanding the long-term perspectives of the topics, the book makes an effort in highlighting their impact as a modern tool for the growth of the discipline. It will serve as a reference to a broad spectrum of readers especially in the fields of chemical engineering, material science and metallurgy.

To facilitate a deeper understanding of the contents of this book a short introduction of every chapter is written below:

Chapter 1- The branch of physical chemistry that examines the relation between electricity and chemical change is known as electrochemistry. The branch of engineering which studies this is known as electrochemical engineering. The chapter on electrochemistry offers an insightful focus, keeping in mind the complex subject matter.

Chapter 2- Electrochemistry is an interdisciplinary subject. It spreads to other fields as well, fields such as photoelectrochemistry, bioelectrochemistry and quantum electrochemistry. The study of the interaction between light and electrochemical systems is termed as photoelectrochemistry. Another branch of electrochemistry that has been discussed is bioelectrochemistry; it concerns itself with topics like cell membrane potentials and electrode reactions of redox enzymes. This section will provide a glimpse of related fields of electrochemistry.

Chapter 3- Electric charge is the property of matter that makes it experience a force when it is placed in an electromagnetic field. The basic elements of electrochemistry are electric charge, electric current, cathode, anode and chemical kinetics. The following chapter focuses on the important elements of electrochemistry, keeping in mind the complex subject matter.

Chapter 4- A number of devices are used in electrochemistry. Some of these devices are electrochemical cells, galvanic cells, thermogalvanic cells, electrolytic cells, fuel cells and photocathodes. An electrochemical cell is a device that is capable of producing electrical energy from chemical reactions whereas galvanic cells are electrochemical cells that develop electrical energy from automatic redox reactions within a cell. This section will provide an integrated understanding of electrochemistry.

Chapter 5- In order to stop iron or steel from rusting a coat of zinc is applied on it. This process of applying zinc is known as galvanization. The process of bonding steel to protect it from corrosion is termed as electrogalvaning. This chapter helps the reader in understanding the various reactions and processes in electrochemistry.

Chapter 6- The theories and laws related to electrochemistry are Debye–Hückel theory, Paschen's law and Faraday's law of electrolysis. Debye-Hückel theory explains the departures from ideality in solutions of electrolytes and plasmas. The topics discussed in the text are of great importance to broaden the existing knowledge on electrochemistry.

Chapter 7- Electrochemistry is best understood in confluence with the major topics listed in the following chapter. Some of these topics are electrode potential, absolute electrode potential, standard electrode potential, molar conductivity and electrochemical potential. This text helps the reader in developing an in-depth understanding of the major aspects of electrochemistry.

Chapter 8- Corrosion is a process that occurs naturally, and helps in the conversion of metal to chemically stable form. It is the slow demolition of materials by chemicals. This section will provide an integrated understanding of corrosion.

Chapter 9- Electrochemistry, as a branch of study, has gone through several changes, and has evolved over a number of decades. In recent times, electrochemistry has become a part of the present study, including research in batteries and fuel cells. The aspects elucidated in this chapter are of vital importance, and provide a better understanding of electrochemistry.

Finally, I would like to thank the entire team involved in the inception of this book for their valuable time and contribution. This book would not have been possible without their efforts. I would also like to thank my friends and family for their constant support.

Editor

1

Introduction to Electrochemistry

The branch of physical chemistry that examines the relation between electricity and chemical change is known as electrochemistry. The branch of engineering which studies this is known as electrochemical engineering. The chapter on electrochemistry offers an insightful focus, keeping in mind the complex subject matter.

Electrochemistry

Electrochemistry is the branch of physical chemistry that studies the relationship between electricity, as a measurable and quantitative phenomenon, and identifiable chemical change, with either electricity considered an outcome of a particular chemical change or vice versa. These reactions involve electric charges moving between electrodes and an electrolyte (or ionic species in a solution). Thus electrochemistry deals with the interaction between electrical energy and chemical change.

English chemist John Daniell (left) and physicist Michael Faraday (right), both credited as founders of electrochemistry today.

When a chemical reaction is caused by an externally supplied current, as in electrolysis, or if an electric current is produced by a spontaneous chemical reaction as in a battery, it is called an *electrochemical* reaction. Chemical reactions where electrons are transferred directly between molecules and/or atoms are called oxidation-reduction or (redox) reactions. In general, electrochemistry describes the overall reactions when individual redox reactions are separate but connected by an external electric circuit and an intervening electrolyte.

History

16th to 18th Century Developments

Understanding of electrical matters began in the sixteenth century. During this century, the English scientist William Gilbert spent 17 years experimenting with magnetism and, to a lesser extent, electricity. For his work on magnets, Gilbert became known as the *"Father of Magnetism."* He discovered various methods for producing and strengthening magnets.

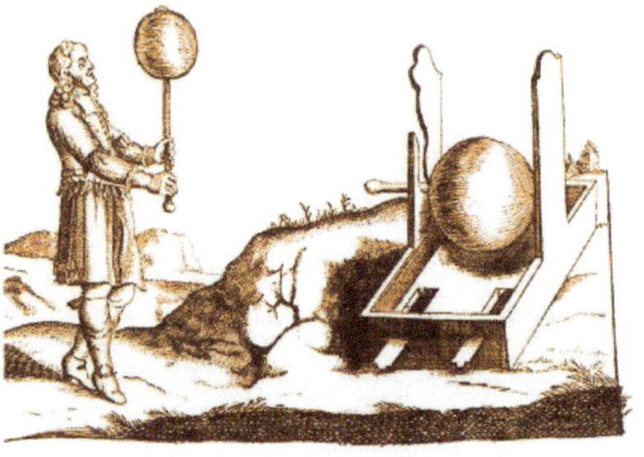

German physicist Otto von Guericke beside his electrical generator while conducting an experiment.

In 1663, the German physicist Otto von Guericke created the first electric generator, which produced static electricity by applying friction in the machine. The generator was made of a large sulfur ball cast inside a glass globe, mounted on a shaft. The ball was rotated by means of a crank and an electric spark was produced when a pad was rubbed against the ball as it rotated. The globe could be removed and used as source for experiments with electricity.

By the mid—18th century the French chemist Charles François de Cisternay du Fay had discovered two types of static electricity, and that like charges repel each other whilst unlike charges attract. Du Fay announced that electricity consisted of two fluids: *"vitreous"* (from the Latin for *"glass"*), or positive, electricity; and *"resinous,"* or negative, electricity. This was the *two-fluid theory* of electricity, which was to be opposed by Benjamin Franklin's *one-fluid theory* later in the century.

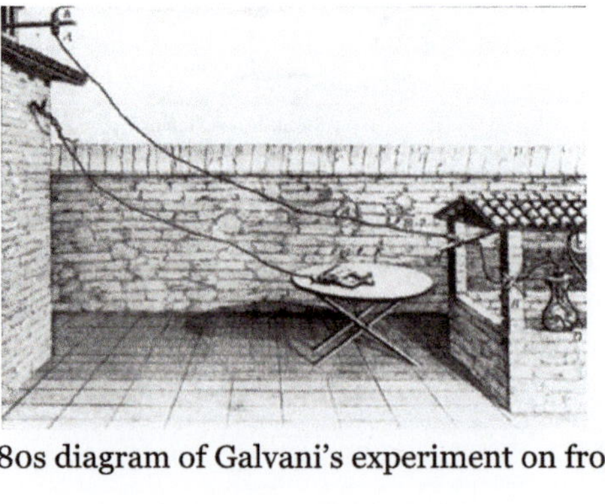

Late 1780s diagram of Galvani's experiment on frog legs.

In 1785, Charles-Augustin de Coulomb developed the law of electrostatic attraction as an outgrowth of his attempt to investigate the law of electrical repulsions as stated by Joseph Priestley in England.

Italian physicist Alessandro Volta showing his *"battery"* to French emperor Napoleon Bonaparte in the early 19th century.

In the late 18th century the Italian physician and anatomist Luigi Galvani marked the birth of electrochemistry by establishing a bridge between chemical reactions and electricity on his essay *"De Viribus Electricitatis in Motu Musculari Commentarius"* (Latin for Commentary on the Effect of Electricity on Muscular Motion) in 1791 where he proposed a *"nerveo-electrical substance"* on biological life forms.

In his essay Galvani concluded that animal tissue contained a here-to-fore neglected innate, vital force, which he termed *"animal electricity,"* which activated nerves and muscles spanned by metal probes. He believed that this new force was a form of electricity in addition to the *"natural"* form produced by lightning or by the electric eel and torpedo ray as well as the *"artificial"* form produced by friction (i.e., static electricity).

Galvani's scientific colleagues generally accepted his views, but Alessandro Volta rejected the idea of an *"animal electric fluid,"* replying that the frog's legs responded to differences in metal temper, composition, and bulk. Galvani refuted this by obtaining muscular action with two pieces of the same material.

19th Century

In 1800, William Nicholson and Johann Wilhelm Ritter succeeded in decomposing water into hydrogen and oxygen by electrolysis. Soon thereafter Ritter discovered the process of electroplating. He also observed that the amount of metal deposited and the amount of oxygen produced during an electrolytic process depended on the distance between the electrodes. By 1801, Ritter observed thermoelectric currents and anticipated the discovery of thermoelectricity by Thomas Johann Seebeck.

Sir Humphry Davy's portrait in the 19th century.

By the 1810s, William Hyde Wollaston made improvements to the galvanic cell. Sir Humphry Davy's work with electrolysis led to the conclusion that the production of electricity in simple electrolytic cells resulted from chemical action and that chemical combination occurred between substances of opposite charge. This work led directly to the isolation of sodium and potassium from their compounds and of the alkaline earth metals from theirs in 1808.

Hans Christian Ørsted's discovery of the magnetic effect of electric currents in 1820 was immediately recognized as an epoch-making advance, although he left further work on electromagnetism to others. André-Marie Ampère quickly repeated Ørsted's experiment, and formulated them mathematically.

In 1821, Estonian-German physicist Thomas Johann Seebeck demonstrated the electrical potential in the juncture points of two dissimilar metals when there is a heat difference between the joints.

In 1827, the German scientist Georg Ohm expressed his law in this famous book *"Die galvanische Kette, mathematisch bearbeitet"* (The Galvanic Circuit Investigated Mathematically) in which he gave his complete theory of electricity.

In 1832, Michael Faraday's experiments led him to state his two laws of electrochemistry. In 1836, John Daniell invented a primary cell which solved the problem of polarization by eliminating hydrogen gas generation at the positive electrode. Later results revealed that alloying the amalgamated zinc with mercury would produce a higher voltage.

William Grove produced the first fuel cell in 1839. In 1846, Wilhelm Weber developed the electrodynamometer. In 1868, Georges Leclanché patented a new cell which eventually became the forerunner to the world's first widely used battery, the zinc carbon cell.

Swedish chemist Svante Arrhenius portrait circa 1880s.

Svante Arrhenius published his thesis in 1884 on *Recherches sur la conductibilité galvanique des électrolytes* (Investigations on the galvanic conductivity of electrolytes). From his results the author concluded that electrolytes, when dissolved in water, become to varying degrees split or dissociated into electrically opposite positive and negative ions.

In 1886, Paul Héroult and Charles M. Hall developed an efficient method (the Hall–Héroult process) to obtain aluminium using electrolysis of molten alumina.

In 1894, Friedrich Ostwald concluded important studies of the conductivity and electrolytic dissociation of organic acids.

German scientist Walther Nernst portrait in the 1910s.

Walther Hermann Nernst developed the theory of the electromotive force of the voltaic cell in 1888. In 1889, he showed how the characteristics of the current produced could be used to calculate the free energy change in the chemical reaction producing the current. He constructed an equation, known as Nernst equation, which related the voltage of a cell to its properties.

In 1898, Fritz Haber showed that definite reduction products can result from electrolytic processes if the potential at the cathode is kept constant. In 1898, he explained the reduction of nitrobenzene in stages at the cathode and this became the model for other similar reduction processes.

20th Century and Recent Developments

In 1902, The Electrochemical Society (ECS) was founded.

In 1909, Robert Andrews Millikan began a series of experiments to determine the electric charge carried by a single electron.

In 1923, Johannes Nicolaus Brønsted and Martin Lowry published essentially the same theory about how acids and bases behave, using an electrochemical basis.

In 1937, Arne Tiselius developed the first sophisticated electrophoretic apparatus. Some years later, he was awarded the 1948 Nobel Prize for his work in protein electrophoresis.

A year later, in 1949, the International Society of Electrochemistry (ISE) was founded.

By the 1960s–1970s quantum electrochemistry was developed by Revaz Dogonadze and his pupils.

Principles

Oxidation and Reduction

The term "redox" stands for reduction-oxidation. It refers to electrochemical processes involving electron transfer to or from a molecule or ion changing its oxidation state. This reaction can occur through the application of an external voltage or through the release of chemical energy. Oxidation and reduction describe the change of oxidation state that takes place in the atoms, ions or molecules involved in an electrochemical reaction. Formally, oxidation state is the hypothetical charge that an atom would have if all bonds to atoms of different elements were 100% ionic. An atom or ion that gives up an electron to another atom or ion has its oxidation state increase, and the recipient of the negatively charged electron has its oxidation state decrease.

For example, when atomic sodium reacts with atomic chlorine, sodium donates one electron and attains an oxidation state of +1. Chlorine accepts the electron and its oxidation state is reduced to −1. The sign of the oxidation state (positive/negative) actually corresponds to the value of each ion's electronic charge. The attraction of the differently charged sodium and chlorine ions is the reason they then form an ionic bond.

The loss of electrons from an atom or molecule is called oxidation, and the gain of electrons is reduction. This can be easily remembered through the use of mnemonic devices. Two of the most popular are *"OIL RIG"* (Oxidation Is Loss, Reduction Is Gain) and *"LEO"* the lion says *"GER"* (Lose Electrons: Oxidation, Gain Electrons: Reduction). Oxidation and reduction always occur in a paired fashion such that one species is oxidized when another is reduced. For cases where electrons are shared (covalent bonds) between atoms with large differences in electronegativity, the electron is assigned to the atom with the largest electronegativity in determining the oxidation state.

The atom or molecule which loses electrons is known as the *reducing agent*, or *reductant*, and the substance which accepts the electrons is called the *oxidizing agent*, or *oxidant*. Thus, the oxidizing agent is always being reduced in a reaction; the reducing agent is always being oxidized. Oxygen is a common oxidizing agent, but not the only one. Despite the name, an oxidation reaction does not necessarily need to involve oxygen. In fact, a fire can be fed by an oxidant other than oxygen; fluorine fires are often unquenchable, as fluorine is an even stronger oxidant (it has a higher electronegativity and thus accepts electrons even better) than oxygen.

For reactions involving oxygen, the gain of oxygen implies the oxidation of the atom or molecule to which the oxygen is added (and the oxygen is reduced). In organic compounds, such as butane or ethanol, the loss of hydrogen implies oxidation of the molecule from which it is lost (and the hydrogen is reduced). This follows because the hydrogen donates its electron in covalent bonds with non-metals but it takes the electron along when it is lost. Conversely, loss of oxygen or gain of hydrogen implies reduction.

Balancing Redox Reactions

Electrochemical reactions in water are better understood by balancing redox reactions using the ion-electron method where H^+, OH^- ion, H_2O and electrons (to compensate the oxidation changes) are added to cell's half-reactions for oxidation and reduction.

Acidic Medium

In acid medium H^+ ions and water are added to half-reactions to balance the overall reaction. For example, when manganese reacts with sodium bismuthate.

Unbalanced reaction: $Mn^{2+}_{(aq)} + NaBiO_{3(s)} \rightarrow Bi^{3+}_{(aq)} + MnO_4^-{}_{(aq)}$

Oxidation: $4\,H_2O_{(l)} + Mn^{2+}_{(aq)} \rightarrow MnO_4^-{}_{(aq)} + 8\,H^+_{(aq)} + 5\,e^-$

Reduction: $2\,e^- + 6\,H^+_{(aq)} + BiO_3^-{}_{(s)} \rightarrow Bi^{3+}_{(aq)} + 3\,H_2O_{(l)}$

Finally, the reaction is balanced by multiplying the number of electrons from the reduction half reaction to oxidation half reaction and vice versa and adding both half reactions, thus solving the equation.

$$8\,H_2O_{(l)} + 2\,Mn^{2+}_{(aq)} \rightarrow 2\,MnO_4^-{}_{(aq)} + 16\,H^+_{(aq)} + 10\,e^-$$

$$10\ e^- + 30\ H^+_{(aq)} + 5\ BiO_3{}^-{}_{(s)} \rightarrow 5\ Bi^{3+}{}_{(aq)} + 15\ H_2O_{(l)}$$

Reaction balanced:

$$14\ H^+_{(aq)} + 2\ Mn^{2+}{}_{(aq)} + 5\ NaBiO_{3(s)} \rightarrow 7\ H_2O_{(l)} + 2\ MnO_4{}^-{}_{(aq)} + 5\ Bi^{3+}{}_{(aq)} + 5\ Na^+_{(aq)}$$

Basic Medium

In basic medium OH^- ions and water are added to half reactions to balance the overall reaction. For example, on reaction between potassium permanganate and sodium sulfite.

Unbalanced reaction: $KMnO_4 + Na_2SO_3 + H_2O \rightarrow MnO_2 + Na_2SO_4 + KOH$

Reduction: $3\ e^- + 2\ H_2O + MnO_4{}^- \rightarrow MnO_2 + 4\ OH^-$

Oxidation: $2\ OH^- + SO_3{}^{2-} \rightarrow SO_4{}^{2-} + H_2O + 2\ e^-$

The same procedure as followed on acid medium by multiplying electrons to opposite half reactions solve the equation thus balancing the overall reaction.

$$6\ e^- + 4\ H_2O + 2\ MnO_4{}^- \rightarrow 2\ MnO_2 + 8\ OH^-$$

$$6\ OH^- + 3\ SO_3{}^{2-} \rightarrow 3\ SO_4{}^{2-} + 3\ H_2O + 6e^-$$

Equation balanced:

$$2\ KMnO_4 + 3\ Na_2SO_3 + H_2O \rightarrow 2\ MnO_2 + 3\ Na_2SO_4 + 2\ KOH$$

Neutral Medium

The same procedure as used on acid medium is applied, for example on balancing using electron ion method to complete combustion of propane.

Unbalanced reaction: $C_3H_8 + O_2 \rightarrow CO_2 + H_2O$

Reduction: $4\ H^+ + O_2 + 4\ e^- \rightarrow 2\ H_2O$

Oxidation: $6\ H_2O + C_3H_8 \rightarrow 3\ CO_2 + 20\ e^- + 20\ H^+$

As in acid and basic medium, electrons which were used to compensate oxidation changes are multiplied to opposite half reactions, thus solving the equation.

$$20\ H^+ + 5\ O_2 + 20\ e^- \rightarrow 10\ H_2O$$

$$6\ H_2O + C_3H_8 \rightarrow 3\ CO_2 + 20\ e^- + 20\ H^+$$

Equation balanced:

$$C_3H_8 + 5\ O_2 \rightarrow 3\ CO_2 + 4\ H_2O$$

Electrochemical Cells

An electrochemical cell is a device that produces an electric current from energy released by a spontaneous redox reaction. This kind of cell includes the Galvanic cell or Voltaic cell, named after Luigi Galvani and Alessandro Volta, both scientists who conducted several experiments on chemical reactions and electric current during the late 18th century.

Electrochemical cells have two conductive electrodes (the anode and the cathode). The anode is defined as the electrode where oxidation occurs and the cathode is the electrode where the reduction takes place. Electrodes can be made from any sufficiently conductive materials, such as metals, semiconductors, graphite, and even conductive polymers. In between these electrodes is the electrolyte, which contains ions that can freely move.

The galvanic cell uses two different metal electrodes, each in an electrolyte where the positively charged ions are the oxidized form of the electrode metal. One electrode will undergo oxidation (the anode) and the other will undergo reduction (the cathode). The metal of the anode will oxidize, going from an oxidation state of 0 (in the solid form) to a positive oxidation state and become an ion. At the cathode, the metal ion in solution will accept one or more electrons from the cathode and the ion's oxidation state is reduced to 0. This forms a solid metal that electrodeposits on the cathode. The two electrodes must be electrically connected to each other, allowing for a flow of electrons that leave the metal of the anode and flow through this connection to the ions at the surface of the cathode. This flow of electrons is an electric current that can be used to do work, such as turn a motor or power a light.

A galvanic cell whose electrodes are zinc and copper submerged in zinc sulfate and copper sulfate, respectively, is known as a Daniell cell.

Half reactions for a Daniell cell are these:

Zinc electrode (anode): $Zn_{(s)} \rightarrow Zn^{2+}_{(aq)} + 2\ e^-$

Copper electrode (cathode): $Cu^{2+}_{(aq)} + 2\ e^- \rightarrow Cu_{(s)}$

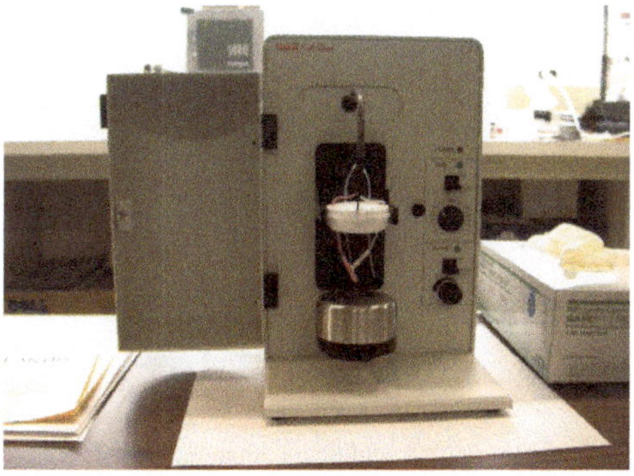

A modern cell stand for electrochemical research. The electrodes attach to high-quality metallic wires, and the stand is attached to a potentiostat/galvanostat (not pictured). A shot glass-shaped container is aerated with a noble gas and sealed with the Teflon block.

In this example, the anode is the zinc metal which is oxidized (loses electrons) to form zinc ions in solution, and copper ions accept electrons from the copper metal electrode and the ions deposit at the copper cathode as an electrodeposit. This cell forms a simple battery as it will spontaneously generate a flow of electric current from the anode to the cathode through the external connection. This reaction can be driven in reverse by applying a voltage, resulting in the deposition of zinc metal at the anode and formation of copper ions at the cathode.

To provide a complete electric circuit, there must also be an ionic conduction path between the anode and cathode electrolytes in addition to the electron conduction path. The simplest ionic conduction path is to provide a liquid junction. To avoid mixing between the two electrolytes, the liquid junction can be provided through a porous plug that allows ion flow while reducing electrolyte mixing. To further minimize mixing of the electrolytes, a salt bridge can be used which consists of an electrolyte saturated gel in an inverted U-tube. As the negatively charged electrons flow in one direction around this circuit, the positively charged metal ions flow in the opposite direction in the electrolyte.

A voltmeter is capable of measuring the change of electrical potential between the anode and the cathode.

Electrochemical cell voltage is also referred to as electromotive force or emf.

A cell diagram can be used to trace the path of the electrons in the electrochemical cell. For example, here is a cell diagram of a Daniell cell:

$$Zn_{(s)} \mid Zn^{2+}\ (1M) \mid\mid Cu^{2+}\ (1M) \mid Cu_{(s)}$$

First, the reduced form of the metal to be oxidized at the anode (Zn) is written. This is separated from its oxidized form by a vertical line, which represents the limit between the phases (oxidation changes). The double vertical lines represent the saline bridge on the cell. Finally, the oxidized form of the metal to be reduced at the cathode, is written, separated from its reduced form by the vertical line. The electrolyte concentration is given as it is an important variable in determining the cell potential.

Standard Electrode Potential

To allow prediction of the cell potential, tabulations of standard electrode potential are available. Such tabulations are referenced to the standard hydrogen electrode (SHE). The standard hydrogen electrode undergoes the reaction

$$2\,H^+_{(aq)} + 2\,e^- \rightarrow H_2$$

which is shown as reduction but, in fact, the SHE can act as either the anode or the cathode, depending on the relative oxidation/reduction potential of the other electrode/electrolyte combination. The term standard in SHE requires a supply of hydrogen gas bubbled through the electrolyte at a pressure of 1 atm and an acidic electrolyte with H^+ activity equal to 1 (usually assumed to be $[H^+] = 1$ mol/liter).

The SHE electrode can be connected to any other electrode by a salt bridge to form a cell. If the second electrode is also at standard conditions, then the measured cell potential is called the standard electrode potential for the electrode. The standard electrode potential for the SHE is zero, by definition. The polarity of the standard electrode potential provides information about the relative reduction potential of the electrode compared to the SHE. If the electrode has a positive potential with respect to the SHE, then that means it is a strongly reducing electrode which forces the SHE to be the anode (an example is Cu in aqueous $CuSO_4$ with a standard electrode potential of 0.337 V). Conversely, if the measured potential is negative, the electrode is more oxidizing than the SHE (such as Zn in $ZnSO_4$ where the standard electrode potential is −0.76 V).

Standard electrode potentials are usually tabulated as reduction potentials. However, the reactions are reversible and the role of a particular electrode in a cell depends on the relative oxidation/reduction potential of both electrodes. The oxidation potential for a particular electrode is just the negative of the reduction potential. A standard cell potential can be determined by looking up the standard electrode potentials for both electrodes (sometimes called half cell potentials). The one that is smaller will be the anode and will undergo oxidation. The cell potential is then calculated as the sum of the reduction potential for the cathode and the oxidation potential for the anode.

$$E^\circ_{cell} = E^\circ_{red}\,(\text{cathode}) - E^\circ_{red}\,(\text{anode}) = E^\circ_{red}\,(\text{cathode}) + E^\circ_{oxi}\,(\text{anode})$$

For example, the standard electrode potential for a copper electrode is:

Cell diagram

$$Pt_{(s)} \mid H_2\,(1\text{ atm}) \mid H^+\,(1\text{ M}) \mid\mid Cu^{2+}\,(1\text{ M}) \mid Cu_{(s)}$$

$$E^\circ_{cell} = E^\circ_{red}\,(\text{cathode}) - E^\circ_{red}\,(\text{anode})$$

At standard temperature, pressure and concentration conditions, the cell's emf (measured by a multimeter) is 0.34 V. By definition, the electrode potential for the SHE is zero. Thus, the Cu is the cathode and the SHE is the anode giving

$$E_{cell} = E^\circ(Cu^{2+}/Cu) - E^\circ(H^+/H_2)$$

Or,

$$E^\circ(Cu^{2+}/Cu) = 0.34\text{ V}$$

Changes in the stoichiometric coefficients of a balanced cell equation will not change E°_{red} value because the standard electrode potential is an intensive property.

Spontaneity of Redox Reaction

During operation of electrochemical cells, chemical energy is transformed into electrical energy and is expressed mathematically as the product of the cell's emf and the electric charge transferred through the external circuit.

$$\text{Electrical energy} = E_{cell}C_{trans}$$

where E_{cell} is the cell potential measured in volts (V) and C_{trans} is the cell current integrated over time and measured in coulombs (C); C_{trans} can also be determined by multiplying the total number of electrons transferred (measured in moles) times Faraday's constant (F).

The emf of the cell at zero current is the maximum possible emf. It is used to calculate the maximum possible electrical energy that could be obtained from a chemical reaction. This energy is referred to as electrical work and is expressed by the following equation:

$$W_{max} = W_{electrical} = -nFE_{cell},$$

where work is defined as positive into the system.

Since the free energy is the maximum amount of work that can be extracted from a system, one can write:

$$\Delta G = -nFE_{cell}$$

A positive cell potential gives a negative change in Gibbs free energy. This is consistent with the cell production of an electric current from the cathode to the anode through the external circuit. If the current is driven in the opposite direction by imposing an external potential, then work is done on the cell to drive electrolysis.

A spontaneous electrochemical reaction (change in Gibbs free energy less than zero) can be used to generate an electric current in electrochemical cells. This is the basis of all batteries and fuel cells. For example, gaseous oxygen (O_2) and hydrogen (H_2) can be combined in a fuel cell to form water and energy, typically a combination of heat and electrical energy.

Conversely, non-spontaneous electrochemical reactions can be driven forward by the application of a current at sufficient voltage. The electrolysis of water into gaseous oxygen and hydrogen is a typical example.

The relation between the equilibrium constant, *K*, and the Gibbs free energy for an electrochemical cell is expressed as follows:

$$\Delta G^{\circ} = -RT\ln K = -nFE^{\circ}_{cell}.$$

Rearranging to express the relation between standard potential and equilibrium constant yields

$$E^{o}_{cell} = \frac{RT}{nF} \ln K \,.$$

The previous equation can use Briggsian logarithm as shown below:

$$E^{o}_{cell} = \frac{0.0591\text{V}}{n} \log K$$

Cell Emf Dependency on Changes in Concentration

Nernst Equation

The standard potential of an electrochemical cell requires standard conditions (ΔG°) for all of the reactants. When reactant concentrations differ from standard conditions, the cell potential will deviate from the standard potential. In the 20th century German chemist Walther Nernst proposed a mathematical model to determine the effect of reactant concentration on electrochemical cell potential.

In the late 19th century, Josiah Willard Gibbs had formulated a theory to predict whether a chemical reaction is spontaneous based on the free energy

$$\Delta G = \Delta G^{\circ} + RT \ln Q$$

Here ΔG is change in Gibbs free energy, ΔG^{o} is the cell potential when Q is equal to 1, T is absolute temperature(Kelvin), R is the gas constant and Q is reaction quotient which can be found by dividing products by reactants using only those products and reactants that are aqueous or gaseous.

Gibbs' key contribution was to formalize the understanding of the effect of reactant concentration on spontaneity.

Based on Gibbs' work, Nernst extended the theory to include the contribution from electric potential on charged species. As shown in the previous section, the change in Gibbs free energy for an electrochemical cell can be related to the cell potential. Thus, Gibbs' theory becomes

$$nF\Delta E = nF\Delta E^{\circ} - RT \ln Q$$

Here n is the number of electrons/mole product, F is the Faraday constant (coulombs/mole), and ΔE is cell potential.

Finally, Nernst divided through by the amount of charge transferred to arrive at a new equation which now bears his name:

$$\Delta E = \Delta E^{\circ} - \frac{RT}{nF} \ln Q$$

Assuming standard conditions (T = 25 °C) and R = 8.3145 J/(K·mol), the equation above can be expressed on base—10 logarithm as shown below:

$$\Delta E = \Delta E^{\circ} - \frac{0.05916\text{V}}{n} \log Q$$

Concentration Cells

A concentration cell is an electrochemical cell where the two electrodes are the same material, the electrolytes on the two half-cells involve the same ions, but the electrolyte concentration differs between the two half-cells.

An example is an electrochemical cell, where two copper electrodes are submerged in two copper(II) sulfate solutions, whose concentrations are 0.05 M and 2.0 M, connected through a salt bridge. This type of cell will generate a potential that can be predicted by the Nernst equation. Both can undergo the same chemistry (although the reaction proceeds in reverse at the anode)

$$Cu^{2+}_{(aq)} + 2\ e^- \rightarrow Cu_{(s)}$$

Le Chatelier's principle indicates that the reaction is more favorable to reduction as the concentration of Cu^{2+} ions increases. Reduction will take place in the cell's compartment where concentration is higher and oxidation will occur on the more dilute side.

The following cell diagram describes the cell mentioned above:

$$Cu_{(s)} \mid Cu^{2+}\ (0.05\ M) \mid\mid Cu^{2+}\ (2.0\ M) \mid Cu_{(s)}$$

Where the half cell reactions for oxidation and reduction are:

Oxidation: $Cu_{(s)} \rightarrow Cu^{2+}$ (0.05 M) + 2 e^-

Reduction: Cu^{2+} (2.0 M) + 2 $e^- \rightarrow Cu_{(s)}$

Overall reaction: Cu^{2+} (2.0 M) $\rightarrow Cu^{2+}$ (0.05 M)

The cell's emf is calculated through Nernst equation as follows:

$$E = E^{\circ} - \frac{0.05916\text{V}}{2} \log \frac{[\text{Cu}^{2+}]_{diluted}}{[\text{Cu}^{2+}]_{concentrated}}$$

The value of E° in this kind of cell is zero, as electrodes and ions are the same in both half-cells.

After replacing values from the case mentioned, it is possible to calculate cell's potential:

$$E = 0 - \frac{0.05916\text{V}}{2}\log\frac{0.05}{2.0} = 0.0474\text{V}$$

or by:

$$E \aleph 0 \quad \frac{0.0257\text{V}}{2}\ln\frac{0.05}{2.0} \quad 0.0474\text{V}$$

However, this value is only approximate, as reaction quotient is defined in terms of ion activities which can be approximated with the concentrations as calculated here.

The Nernst equation plays an important role in understanding electrical effects in cells and organelles. Such effects include nerve synapses and cardiac beat as well as the resting potential of a somatic cell.

Battery

Many types of battery have been commercialized and represent an important practical application of electrochemistry. Early wet cells powered the first telegraph and telephone systems, and were the source of current for electroplating. The zinc-manganese dioxide dry cell was the first portable, non-spillable battery type that made flashlights and other portable devices practical. The mercury battery using zinc and mercuric oxide provided higher levels of power and capacity than the original dry cell for early electronic devices, but has been phased out of common use due to the danger of mercury pollution from discarded cells.

The lead–acid battery was the first practical secondary (rechargeable) battery that could have its capacity replenished from an external source. The electrochemical reaction that produced current was (to a useful degree) reversible, allowing electrical energy and chemical energy to be interchanged as needed. Common lead acid batteries contain a mixture of acid and water, as well as lead plates. The most common mixture used today is 30% acid. One problem however is if left uncharged acid will crystallize within the lead plates of the battery rendering it useless. These batteries last an average of 3 years with daily use however it is not unheard of for a lead acid battery to still be functional after 7–10 years. Lead-acid cells continue to be widely used in automobiles.

All the preceding types have water-based electrolytes, which limits the maximum voltage per cell. The freezing of water limits low temperature performance. The lithium battery, which does not (and cannot) use water in the electrolyte, provides improved performance over other types; a rechargeable lithium ion battery is an essential part of many mobile devices.

The flow battery, an experimental type, offers the option of vastly larger energy capacity because its reactants can be replenished from external reservoirs. The fuel cell can turn the chemical energy bound in hydrocarbon gases or hydrogen directly into electrical energy with much higher efficiency than any combustion process; such devices have powered many spacecraft and are being applied to grid energy storage for the public power system.

Corrosion

Corrosion is an electrochemical process, which reveals itself in rust or tarnish on metals like iron or copper and their respective alloys, steel and brass.

Iron Corrosion

For iron rust to occur the metal has to be in contact with oxygen and water, although chemical reactions for this process are relatively complex and not all of them are completely understood. It is believed the causes are the following: Electron transfer (reduction-oxidation)

One area on the surface of the metal acts as the anode, which is where the oxidation (corrosion) occurs. At the anode, the metal gives up electrons.

$$Fe_{(s)} \rightarrow Fe^{2+}{}_{(aq)} + 2\ e^-$$

Electrons are transferred from iron, reducing oxygen in the atmosphere into water on the cathode, which is placed in another region of the metal.

$$O_{2(g)} + 4\ H^+{}_{(aq)} + 4\ e^- \rightarrow 2\ H_2O_{(l)}$$

Global reaction for the process:

$$2\ Fe_{(s)} + O_{2(g)} + 4\ H^+{}_{(aq)} \rightarrow 2\ Fe^{2+}{}_{(aq)} + 2\ H_2O_{(l)}$$

Standard emf for iron rusting:

$$E^\circ = E^\circ\ (\text{cathode}) - E^\circ\ (\text{anode})$$

$$E^\circ = 1.23V - (-0.44\ V) = 1.67\ V$$

Iron corrosion takes place in an acid medium; H^+ ions come from reaction between carbon dioxide in the atmosphere and water, forming carbonic acid. Fe^{2+} ions oxidizes, following this equation:

$$4Fe^2+_{(aq)} + O2_{(g)} + (4+2x)H2O_{(l)} -> 2Fe2O3.xH2O + 8H+_{(aq)}$$

Iron(III) oxide hydrate is known as rust. The concentration of water associated with iron oxide varies, thus the chemical formula is represented by $2Fe2O3.xH2O$..

An electric circuit is formed as passage of electrons and ions occurs, thus if an electrolyte is present it will facilitate oxidation, explaining why rusting is quicker in salt water.

Corrosion of Common Metals

Coinage metals, such as copper and silver, slowly corrode through use. A patina of green-blue copper carbonate forms on the surface of copper with exposure to the water and carbon dioxide in the air. Silver coins or cutlery that are exposed to high sulfur foods such as eggs or the low levels of sulfur species in the air develop a layer of black Silver sulfide.

Gold and platinum are extremely difficult to oxidize under normal circumstances, and require exposure to a powerful chemical oxidizing agent such as aqua regia.

Some common metals oxidize extremely rapidly in air. Titanium and aluminium oxidize instantaneously in contact with the oxygen in the air. These metals form an extremely thin layer of oxidized metal on the surface. This thin layer of oxide protects the underlying layers of the metal from the air preventing the entire metal from oxidizing. These metals are used in applications where corrosion resistance is important. Iron, in contrast, has an oxide that forms in air and water, called rust, that does not stop the further oxidation of the iron. Thus iron left exposed to air and water will continue to rust until all of the iron is oxided.

Prevention of Corrosion

Attempts to save a metal from becoming anodic are of two general types. Anodic regions dissolve and destroy the structural integrity of the metal.

While it is almost impossible to prevent anode/cathode formation, if a non-conducting material covers the metal, contact with the electrolyte is not possible and corrosion will not occur.

Coating

Metals can be coated with paint or other less conductive metals (*passivation*). This prevents the metal surface from being exposed to electrolytes. Scratches exposing the metal substrate will result in corrosion. The region under the coating adjacent to the scratch acts as the anode of the reaction.

Sacrificial Anodes

A method commonly used to protect a structural metal is to attach a metal which is more anodic than the metal to be protected. This forces the structural metal to be cathodic, thus spared corrosion. It is called *"sacrificial"* because the anode dissolves and has to be replaced periodically.

Zinc bars are attached to various locations on steel ship hulls to render the ship hull cathodic. The zinc bars are replaced periodically. Other metals, such as magnesium, would work very well but zinc is the least expensive useful metal.

To protect pipelines, an ingot of buried or exposed magnesium (or zinc) is buried beside the pipeline and is connected electrically to the pipe above ground. The pipeline is forced to be a cathode and is protected from being oxidized and rusting. The magnesium anode is sacrificed. At intervals new ingots are buried to replace those lost.

Electrolysis

The spontaneous redox reactions of a conventional battery produce electricity through the different chemical potentials of the cathode and anode in the electrolyte. However, electrolysis requires an external source of electrical energy to induce a chemical reaction, and this process takes place in a compartment called an electrolytic cell.

Electrolysis of Molten Sodium Chloride

When molten, the salt sodium chloride can be electrolyzed to yield metallic sodium and gaseous chlorine. Industrially this process takes place in a special cell named Down's cell. The cell is connected to an electrical power supply, allowing electrons to migrate from the power supply to the electrolytic cell.

Reactions that take place at Down's cell are the following:

Anode (oxidation): $2\ Cl^- \rightarrow Cl_{2(g)} + 2\ e^-$

Cathode (reduction): $2\ Na^+_{(l)} + 2\ e^- \rightarrow 2\ Na_{(l)}$

Overall reaction: $2\ Na^+ + 2\ Cl^-_{(l)} \rightarrow 2\ Na_{(l)} + Cl_{2(g)}$

This process can yield large amounts of metallic sodium and gaseous chlorine, and is widely used on mineral dressing and metallurgy industries.

The emf for this process is approximately −4 V indicating a (very) non-spontaneous process. In order for this reaction to occur the power supply should provide at least a potential of 4 V. However, larger voltages must be used for this reaction to occur at a high rate.

Electrolysis of Water

Water can be converted to its component elemental gasses, H_2 and O_2 through the application of an external voltage. Water doesn't decompose into hydrogen and oxygen spontaneously as the Gibbs free energy for the process at standard conditions is about 474.4 kJ. The decomposition of water into hydrogen and oxygen can be performed in an electrolytic cell. In it, a pair of inert electrodes usually made of platinum immersed in water act as anode and cathode in the electrolytic process. The electrolysis starts with the application of an external voltage between the electrodes. This process will not occur except at extremely high voltages without an electrolyte such as sodium chloride or sulfuric acid (most used 0.1 M).

Bubbles from the gases will be seen near both electrodes. The following half reactions describe the process mentioned above:

Anode (oxidation): $2\ H_2O_{(l)} \rightarrow O_{2(g)} + 4\ H^+_{(aq)} + 4\ e^-$

Cathode (reduction): $2\ H_2O_{(g)} + 2\ e^- \rightarrow H_{2(g)} + 2\ OH^-_{(aq)}$

Overall reaction: $2\ H_2O_{(l)} \rightarrow 2\ H_{2(g)} + O_{2(g)}$

Although strong acids may be used in the apparatus, the reaction will not net consume the acid. While this reaction will work at any conductive electrode at a sufficiently large potential, platinum catalyzes both hydrogen and oxygen formation, allowing for relatively mild voltages (~2 V depending on the pH).

Electrolysis of Aqueous Solutions

Electrolysis in an aqueous is a similar process as mentioned in electrolysis of water. However, it is considered to be a complex process because the contents in solution have to be analyzed in half reactions, whether reduced or oxidized.

Electrolysis of a Solution of Sodium Chloride

The presence of water in a solution of sodium chloride must be examined in respect to its reduction and oxidation in both electrodes. Usually, water is electrolysed as mentioned in electrolysis of water yielding *gaseous oxygen in the anode* and gaseous hydrogen in the cathode. On the other hand, sodium chloride in water dissociates in Na^+ and Cl^- ions, cation, which is the positive ion, will be attracted to the cathode (-), thus reducing the sodium ion. The anion will then be attracted to the anode (+) oxidizing chloride ion.

The following half reactions describes the process mentioned:

1. Cathode: $Na^+_{(aq)} + e^- \rightarrow Na_{(s)}$ $\quad E^o_{red} = -2.71\ V$

2. Anode: $2\ Cl^-_{(aq)} \rightarrow Cl_{2(g)} + 2\ e^-$ $\quad E^o_{red} = +1.36\ V$

3. Cathode: $2\ H_2O_{(l)} + 2\ e^- \rightarrow H_{2(g)} + 2\ OH^-_{(aq)}$ $\quad E^o_{red} = -0.83\ V$

4. Anode: $2\ H_2O_{(l)} \rightarrow O_{2(g)} + 4\ H^+_{(aq)} + 4\ e^-$ $\quad E^o_{red} = +1.23\ V$

Reaction 1 is discarded as it has the most negative value on standard reduction potential thus making it less thermodynamically favorable in the process.

When comparing the reduction potentials in reactions 2 and 4, the reduction of chloride ion is favored. Thus, if the Cl^- ion is favored for reduction, then the water reaction is favored for oxidation producing gaseous oxygen, however experiments show gaseous chlorine is produced and not oxygen.

Although the initial analysis is correct, there is another effect that can happen, known as the overvoltage effect. Additional voltage is sometimes required, beyond the voltage predicted by the $E°_{cell}$. This may be due to kinetic rather than thermodynamic considerations. In fact, it has been proven that the activation energy for the chloride ion is very low, hence favorable in kinetic terms. In other words, although the voltage applied is thermodynamically sufficient to drive electrolysis, the rate is so slow that to make the process proceed in a reasonable time frame, the voltage of the external source has to be increased (hence, overvoltage).

Finally, reaction 3 is favorable because it describes the proliferation of OH^- ions thus letting a probable reduction of H^+ ions less favorable an option.

The overall reaction for the process according to the analysis would be the following:

Anode (oxidation): $2\ Cl^-_{(aq)} \rightarrow Cl_{2(g)} + 2\ e^-$

Cathode (reduction): $2\ H_2O_{(l)} + 2\ e^- \rightarrow H_{2(g)} + 2\ OH^-_{(aq)}$

Overall reaction: $2\ H_2O + 2\ Cl^-_{(aq)} \rightarrow H_{2(g)} + Cl_{2(g)} + 2\ OH^-_{(aq)}$

As the overall reaction indicates, the concentration of chloride ions is reduced in comparison to OH^- ions (whose concentration increases). The reaction also shows the production of gaseous hydrogen, chlorine and aqueous sodium hydroxide.

Quantitative Electrolysis and Faraday's Laws

Quantitative aspects of electrolysis were originally developed by Michael Faraday in 1834. Faraday is also credited to have coined the terms *electrolyte*, electrolysis, among many others while he studied quantitative analysis of electrochemical reactions. Also he was an advocate of the law of conservation of energy.

First Law

Faraday concluded after several experiments on electric current in non-spontaneous process, the mass of the products yielded on the electrodes was proportional to the value of current supplied to the cell, the length of time the current existed, and the molar mass of the substance analyzed. In other words, the amount of a substance deposited on each electrode of an electrolytic cell is directly proportional to the quantity of electricity passed through the cell.

Below is a simplified equation of Faraday's first law:

$$m = \frac{1}{96485(\mathrm{C \cdot mol^{-1}})} \cdot \frac{QM}{n}$$

Where

> m is the mass of the substance produced at the electrode (in grams),
>
> Q is the total electric charge that passed through the solution (in coulombs),
>
> n is the valence number of the substance as an ion in solution (electrons per ion),
>
> M is the molar mass of the substance (in grams per mole).

Second Law

Faraday devised the laws of chemical electrodeposition of metals from solutions in 1857. He formulated the second law of electrolysis stating *"the amounts of bodies which are equivalent to each other in their ordinary chemical action have equal quantities of electricity naturally associated with them."* In other words, the quantities of different elements deposited by a given amount of electricity are in the ratio of their chemical equivalent weights.

An important aspect of the second law of electrolysis is electroplating which together with the first law of electrolysis, has a significant number of applications in the industry, as when used to protect metals to avoid corrosion.

Applications

There are various extremely important electrochemical processes in both nature and industry, like the coating of objects with metals or metal oxides through electrodeposition and the detection of alcohol in drunken drivers through the redox reaction of ethanol. The generation of chemical energy through photosynthesis is inherently an electrochemical process, as is production of metals like aluminum and titanium from their ores. Certain diabetes blood sugar meters measure the amount of glucose in the blood through its redox potential. As well as the established electrochemical technologies (like deep cycle lead acid batteries) there is also a wide range of new emerging technologies such as fuel cells, large format lithium ion batteries, electrochemical reactors and super-capacitors that are becoming increasingly commercial. Electrochemistry has also important applications in the food industry,like the assessment of food/package interactions, the analysis of milk composition, the characterization and the determination of the freezing end-point of ice-cream mixes, the determination of free acidity in olive oil.

The action potentials that travel down neurons are based on electric current generated by the movement of sodium and potassium ions into and out of cells. Specialized cells in certain animals like the electric eel can generate electric currents powerful enough to disable much larger animals.

Electrochemical Engineering

Electrochemical engineering is the branch of engineering dealing with the technological applications of electrochemical phenomena, such as synthesis of chemicals, electrowinning and refining of metals, batteries and fuel cells, sensors, surface modification by electrodeposition and etching, separations, and corrosion. It is an overlap between electrochemistry and chemical engineering. One of the pioneers of this field of engineering was Charles Frederick Burgess.

More than 6% of the electrical energy is consumed by electrochemical operations in the USA.

History

This branch of engineering emerged gradually from chemical engineering. The works of Wagner (1962) and Levich (1962) influenced the emergence of electrochemical engineering, because their work inspired so many others. Several individuals, including Tobias, Ibl, and Hine, established engineering training centers and, with their colleagues, developed important experimental and theoretical methods of study.

Scope

Electrochemical engineering combines the study of heterogeneous charge transfer at electrode/electrolyte interphases with the development of practical materials and processes. Fundamental considerations include electrode materials and the kinetics of redox species. The development of the technology involves the study of the electrochemical reactors, their potential and current distribution, mass transport conditions, hydrodynamics, geometry and components as well as the quantification of its overall performance in terms of reaction yield, conversion efficiency, and energy efficiency. Industrial developments require further reactor and process design, fabrication methods, testing and product development.

Applications

Electrochemical engineering finds applications in chemical synthesis, ion/organics removal, deposition of fils of metals and semiconductors, sensors and monitoring and energy storage and conversion.

References

- Richard P. Olenick, Tom M. Apostol, David L. Goodstein Beyond the mechanical universe: from electricity to modern physics, Cambridge University Press (1986) ISBN 0-521-30430-X, p. 160
- J. A. M. Bleeker, Johannes Geiss, M. Huber The century of space science, Volume 1, Springer (2001) ISBN 0-7923-7196-8 p. 227

- John Robert Norris, Douglas W. Ribbons (1972) Methods in microbiology, Volume 6, Academic Press. ISBN 0-12-521546-0 p. 248
- William Berkson (1974) Fields of force: the development of a world view from Faraday to Einstein, Routledge. ISBN 0-7100-7626-6 pp. 34 ff
- Brian Scott Baigrie Electricity and magnetism: a historical perspective, Greenwood Publishing Group (2007) ISBN 0-313-33358-0 p. 73
- Polmear, I.J. (2006). "Production of Aluminium". Light alloys from traditional alloys to nanocrystals. Oxford: Elsevier/Butterworth-Heinemann. pp. 15–16. ISBN 978-0-7506-6371-7.
- William L. Masterton, Cecile N. Hurley Chemistry: Principles and Reactions, Cengage Learning (2008) ISBN 0-495-12671-3 p. 379

2

Branches of Electrochemistry

Electrochemistry is an interdisciplinary subject. It spreads to other fields as well, fields such as photoelectrochemistry, bioelectrochemistry and quantum electrochemistry. The study of the interaction between light and electrochemical systems is termed as photoelectrochemistry. Another branch of electrochemistry that has been discussed is bioelectrochemistry; it concerns itself with topics like cell membrane potentials and electrode reactions of redox enzymes. This section will provide a glimpse of related fields of electrochemistry.

Photoelectrochemistry

Photoelectrochemistry is a subfield of study within physical chemistry concerned with the interaction of light with electrochemical systems. It is an active domain of investigation. One of the pioneers of this field of electrochemistry was the German electrochemist Heinz Gerischer. The interest in this domain is high in the context of development of renewable energy conversion and storage technology.

Historical Approach

Photoelectrochemistry has been intensively studied in the 70-80s because of the first peak oil crisis. Because fossil fuels are non-renewable, it is necessary to develop processes to obtain renewable ressources and use clean energy. Artificial photosynthesis, photoelectrochemical water splitting and regenerative solar cells are of special interest in this context.

H. Gerischer, H. Tributsch, AJ. Nozik, AJ. Bard, A. Fujishima, K. Honda, PE. Laibinis, K. Rajeshwar, TJ Meyer, PV. Kamat, N.S. Lewis, R. Memming, JOM. Bockris are researchers which have contributed a lot to the field of photoelectrochemistry.

Semiconductor's Electrochemistry

Introduction

Semiconductor material has a band gap and generates a pair of electron and hole per absorbed photon if the energy of the photon is higher than the band gap of the semiconductor. This property of semiconductor materials has been successfully used to convert solar energy into electrical energy by photovoltaic devices.

In photocatalysis the electron-hole pair is immediately used to drive a redox reaction but the problem is that the electron-hole pair suffer from fast recombinations. In photoelectrocatalysis, a differential potential is applied to diminish the number of recombinations between the electrons and the holes. This allows an increase in the yield of light's conversion into chemical energy.

Semiconductor-electrolyte Interface

When a semiconductor comes into contact with a liquid (redox species), to maintain electrostatic equilibrium, there will be a charge transfer between the semiconductor and liquid phase if formal redox potential of redox species lies inside semiconductor band gap. At thermodynamic equilibrium, the Fermi level of semiconductor and the formal redox potential of redox species are aligned at the interface between semiconductor and redox species. This introduces a downward band bending in a n-type semiconductor for n-type semiconductor/liquid junction (Figure 1(a)) and an upward band bending in a p-type semiconductor for a p-type semiconductor/liquid junction (Figure 1(b)). This characteristic of semiconductor/liquid junctions is similar to a rectifying semiconductor/metal junction or Schottky junction. Ideally to get a good rectifying characteristics at the semiconductor/liquid interface, the formal redox potential must be close to the valence band of the semiconductor for a n-type semiconductor and close to the conduction band of the semiconductor for a p-type semiconductor. The semiconductor/liquid junction has one advantage over the rectifying semiconductor/metal junction in that the light is able to travel through to the semiconductor surface without much reflection; whereas most of the light is reflected back from the metal surface at a semiconductor/metal junction. Therefore, semiconductor/liquid junctions can also be used as photovoltaic devices similar to solid state p–n junction devices. Both n-type and p-type semiconductor/liquid junctions can be used as photovoltaic devices to convert solar energy into electrical energy and are called photoelectrochemical cells. In addition, a semiconductor/liquid junction could also be used to directly convert solar energy into chemical energy by virtue of photoelectrolysis at the semiconductor/liquid junction.

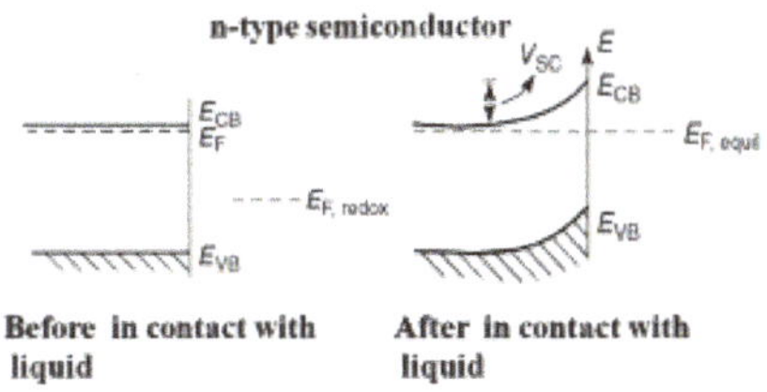

Figure 1(a) band diagram of n-type semiconductor/liquid junction

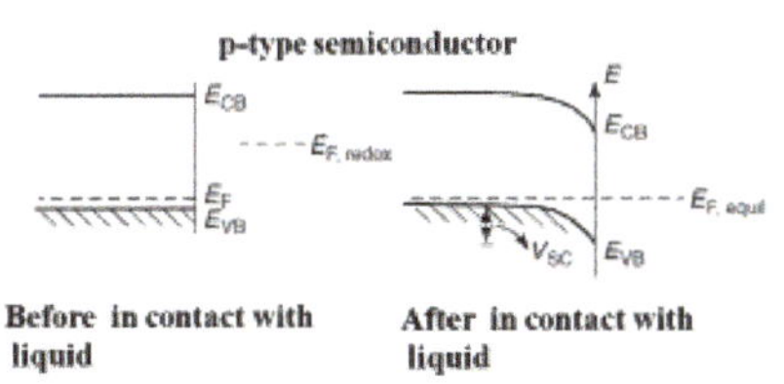

Figure 1(b) band diagram of p-type semiconductor/liquid junction

Experimental Setup

Semiconductors are usually studied in a photoelectrochemical cell. Different configurations exist with a three electrode device. The phenomenon to study happens at the working electrode WE while the differential potential is applied between the WE and a reference electrode RE (saturated calomel, Ag/AgCl). The current is measured between the WE and the counter electrode CE (carbon vitreous, platinum gauze). The working electrode is the semiconductor material and the electrolyte is composed of a solvent, an electrolyte and a redox specie.

A UV-vis lamp is usually used to illuminate the working electrode. The photoelectrochemical cell is usually made with a quartz window because it does not absorb the light. A monochromator can be used to control the wavelength sent to the WE.

Main Absorbers Used in Photoelectrochemistry

Semiconductor IV

C(diamond), Si, Ge, SiC, SiGe

Semiconductor III-V

BN, BP, BAs, AlN, AlP, AlAs, GaN, GaP, GaAs, InN, InP, InAs...

Semiconductor II-VI

CdS, CdSe, CdTe, ZnO, ZnS, ZnSe, ZnTe, MoS_2, $MoSe_2$, $MoTe_2$, WS_2, WSe_2

Metal Oxides

TiO_2, Fe_2O_3, Cu_2O

Organic Dyes

Methylene blue...

Organometallic Dyes

Perovskites

Applications

Photoelectrochemical Splitting of Water

Photoelectrochemistry has been intensively studied in the field of hydrogen production from water and solar energy. The photoelectrochemical splitting of water was historically discovered by Fujishima and Honda in 1972 onto TiO_2 electrodes. Recently many materials have shown promising properties to split efficiently water but TiO_2 remains cheap, abundant, stable against photo-corrosion. The main problem of TiO_2 is its band-gap which is 3 or 3.2 eV according to its crystallinity (anatase or rutile). These values are too high and only the wavelength in the UV region can be absorbed. To increase the performances of this material to split water with solar wavelength, it is necessary to sensitize the TiO_2. Currently Quantum Dots sensitization is very promising but more research is needed to find new materials able to absorb the light efficiently. Recently water a splitting membrane concept has been developed. This method to split water is very similar to the principle of fuel cells but in a reverse way.

Artificial Photosynthesis

Photosynthesis is the natural process that converts CO_2 using light to produce hydro-carbon compounds such as sugar. The depletion of fossil fuels encourages scientists to find alternatives to produce hydrocarbon compounds. Artificial photosynthesis is a promising method mimicking the natural photosynthesis to produce such compounds. The photoelectrochemical reduction of CO2 is much studied because of its worldwide impact. Many researchers aim to find new semiconductors to develop stable and efficient photo-anodes and photo-cathodes.

Regenerative Cells or Dye-sensitized Solar Cell (Graetzel Cell)

Dye-sensitized solar cells or DSSCs use TiO_2 and dyes to absorb the light. This absorption induces the formation of electron-hole pairs which are used to oxidize and reduce the same redox couple, usually I^-/I_3^-. Consequently, a differential potential is created which induces a current.

Bioelectrochemistry

Bioelectrochemistry is a branch of electrochemistry and biophysical chemistry concerned with electrophysiological topics like cell electron-proton transport, cell membrane potentials and electrode reactions of redox enzymes.

History

The beginnings of bioelectrochemistry, as well as those of electrochemistry, are closely related to physiology through the works of Luigi Galvani and then Alessandro Volta. The first modern work in this field is considered that of the German physiologist Julius Bernstein (1902) concerning the source of biopotentials due to different ion concentration through the cell's membrane. The domain of bioelectrochemistry has grown considerably over the past century, maintaining the close connections to various medical and biological and engineering disciplines like electrophysiology and biomedical engineering. The achievements in this field have been awarded several Nobel prizes for Physiology or Medicine. Among prominent electrochemists who have contributed to this field one could mention John Bockris.

Quantum Electrochemistry

The scientific school of Quantum electrochemistry began to form in the 1960s under Revaz Dogonadze. Generally speaking, the field comprises the notions arising in electrodynamics, quantum mechanics, and electrochemistry; and so is studied by a very large array of different professional researchers. The fields they reside in include, chemical, electrical and mechanical engineering, chemistry and physics.

More specifically, quantum electrochemistry is the application of quantum mechanical tools such as density functional theory to the study of electrochemical processes, including electron transfer at electrodes. It also includes models such as Marcus theory.

History and Contributors

The first development of "quantum electrochemistry" is somewhat difficult to pin down. This is not very surprising, since the development of quantum mechanics to chemistry can be summarized as the application of quantum wave theory models to atoms and molecules. This being the case, electrochemistry, which is particularly concerned with the electronic states of some particular system, is already, by its nature, tied into the quantum mechanical model of the electron in quantum chemistry. There were proponents of quantum electrochemistry, who applied quantum mechanics to electrochemistry with unusual zeal, clarity and precision. Among them were Revaz Dogonadze and his co-workers. They developed one of the early quantum mechanical models for proton transfer reactions in chemical systems. Dogonadze is a particularly celebrated promoter of quantum electrochemistry, and is also credited with forming an international summer school of quantum electrochemistry centered in Yugoslavia. He was a main author of the *Quantum-Mechanical Theory of Kinetics of the Elementary Act of Chemical, Electrochemical and Biochemical Processes in Polar Liquids*. Another important contributor is Rudolph A. Marcus, who won the Nobel Prize in Chemistry in 1992 for his *Theory of Electron Transfer Reactions in Chemical Systems*.

References

- J.O'M. Bockris, Shahed U.M. Khan, "Quantum Electrochemistry", Plenum Press, New York, 1979, 538 pp. (ISBN 0-306-31143-7)
- "Encyclopedia of Electrochemistry", Vol. 2, Interfacial Kinetics and Mass Transport, Wiley Publishers, 2003, 563 pp. (ISBN 3-527-30394-4).

- J.O'M. Bockris, Shahed U.M. Khan, "Quantum Electrochemistry", Plenum Press, New York, 1979, 538 pp. (ISBN 0-306-31143-7)
- "Encyclopedia of Electrochemistry", Vol. 2, Interfacial Kinetics and Mass Transport, Wiley Publishers, 2003, 563 pp. (ISBN 3-527-30394-4).

3

Basic Elements of Electrochemistry

Electric charge is the property of matter that makes it experience a force when it is placed in an electromagnetic field. The basic elements of electrochemistry are electric charge, electric current, cathode, anode and chemical kinetics. The following chapter focuses on the important elements of electrochemistry, keeping in mind the complex subject matter.

Electric Charge

Electric charge is the physical property of matter that causes it to experience a force when placed in an electromagnetic field. There are two types of electric charges: positive and negative. Like charges repel and unlike attract. An object is negatively charged if it has an excess of electrons, and is otherwise positively charged or uncharged. The SI derived unit of electric charge is the coulomb (C). In electrical engineering, it is also common to use the ampere-hour (Ah), and, in chemistry, it is common to use the elementary charge (e) as a unit. The symbol Q often denotes charge. Early knowledge of how charged substances interact is now called classical electrodynamics, and is still accurate for problems that don't require consideration of quantum effects.

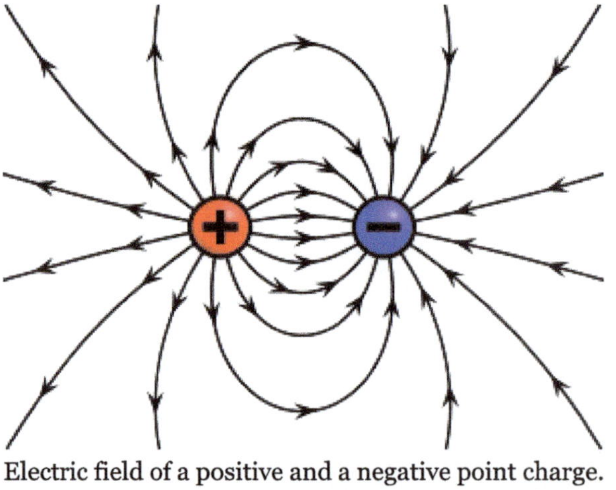

Electric field of a positive and a negative point charge.

The *electric charge* is a fundamental conserved property of some subatomic particles, which determines their electromagnetic interaction. Electrically charged matter is influenced by, and produces, electromagnetic fields. The interaction between a moving charge and an electromagnetic field is the source of the electromagnetic force, which is one of the four fundamental forces.

Twentieth-century experiments demonstrated that electric charge is *quantized*; that is, it comes in integer multiples of individual small units called the elementary charge, *e*, approximately equal to 1.602×10^{-19} coulombs (except for particles called quarks, which have charges that are integer multiples of $\frac{1}{3}e$). The proton has a charge of $+e$, and the electron has a charge of $-e$. The study of charged particles, and how their interactions are mediated by photons, is called quantum electrodynamics.

Overview

Charge is the fundamental property of forms of matter that exhibit electrostatic attraction or repulsion in the presence of other matter. Electric charge is a characteristic property of many subatomic particles. The charges of free-standing particles are integer multiples of the elementary charge *e*; we say that electric charge is *quantized*. Michael Faraday, in his electrolysis experiments, was the first to note the discrete nature of electric charge. Robert Millikan's oil drop experiment demonstrated this fact directly, and measured the elementary charge.

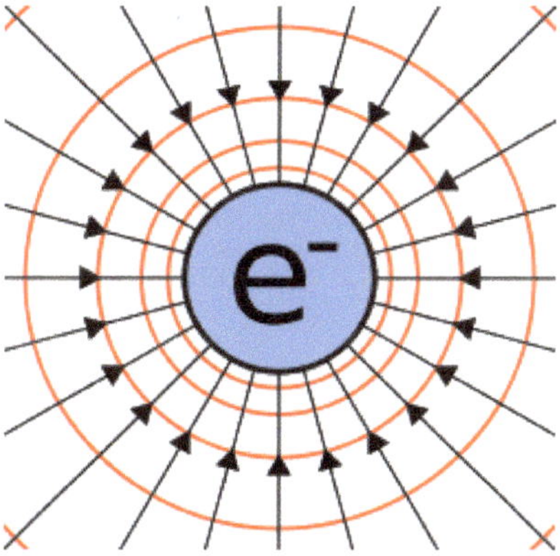

Diagram showing field lines and equipotentials around an electron, a negatively charged particle. In an electrically neutral atom, the number of electrons is equal to the number of protons (which are positively charged), resulting in a net zero overall charge

By convention, the charge of an electron is −1, while that of a proton is +1. Charged particles whose charges have the same sign repel one another, and particles whose charges have different signs attract. Coulomb's law quantifies the electrostatic force between two particles by asserting that the force is proportional to the product of their charges, and inversely proportional to the square of the distance between them.

The charge of an antiparticle equals that of the corresponding particle, but with opposite sign. Quarks have fractional charges of either −1/3 or +2/3, but free-standing quarks have never been observed (the theoretical reason for this fact is asymptotic freedom).

The electric charge of a macroscopic object is the sum of the electric charges of the particles that make it up. This charge is often small, because matter is made of atoms, and atoms typically have equal numbers of protons and electrons, in which case their charges cancel out, yielding a net charge of zero, thus making the atom neutral.

An *ion* is an atom (or group of atoms) that has lost one or more electrons, giving it a net positive charge (cation), or that has gained one or more electrons, giving it a net negative charge (anion). *Monatomic ions* are formed from single atoms, while *polyatomic ions* are formed from two or more atoms that have been bonded together, in each case yielding an ion with a positive or negative net charge.

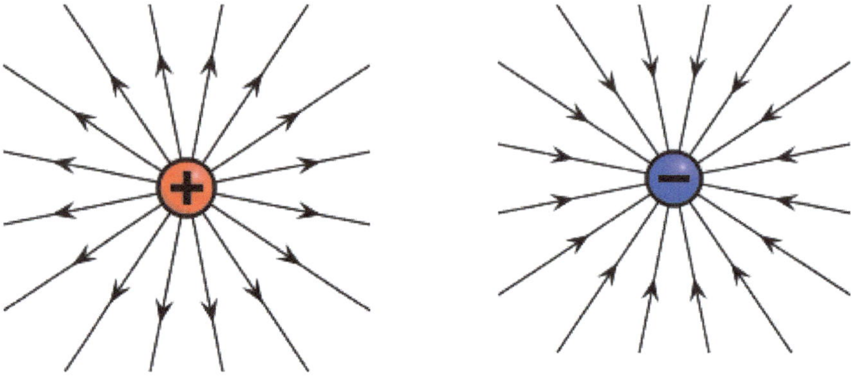

Electric field induced by a positive electric charge (left) and a field induced by a negative electric charge (right).

During formation of macroscopic objects, constituent atoms and ions usually combine to form structures composed of neutral *ionic compounds* electrically bound to neutral atoms. Thus macroscopic objects tend toward being neutral overall, but macroscopic objects are rarely perfectly net neutral.

Sometimes macroscopic objects contain ions distributed throughout the material, rigidly bound in place, giving an overall net positive or negative charge to the object. Also, macroscopic objects made of conductive elements, can more or less easily (depending on the element) take on or give off electrons, and then maintain a net negative or positive charge indefinitely. When the net electric charge of an object is non-zero and motionless, the phenomenon is known as static electricity. This can easily be produced by rubbing two dissimilar materials together, such as rubbing amber with fur or glass with silk. In this way non-conductive materials can be charged to a significant degree, either positively or negatively. Charge taken from one material is moved to the other material, leaving an opposite charge of the same magnitude behind. The law of *conservation of charge* always applies, giving the object from which a negative charge is taken a positive charge of the same magnitude, and vice versa.

Even when an object's net charge is zero, charge can be distributed non-uniformly in the object (e.g., due to an external electromagnetic field, or bound polar molecules). In such cases the object is said to be polarized. The charge due to polarization is known as bound charge, while charge on an object produced by electrons gained or lost from outside the object is called *free charge*. The motion of electrons in conductive metals in a specific direction is known as electric current.

Units

The SI unit of quantity of electric charge is the coulomb, which is equivalent to about 6.242×10^{18} e (e is the charge of a proton). Hence, the charge of an electron is approximately -1.602×10^{-19} C. The coulomb is defined as the quantity of charge that has passed through the cross section of an electrical conductor carrying one ampere within one second. The symbol Q is often used to denote a quantity of electricity or charge. The quantity of electric charge can be directly measured with an electrometer, or indirectly measured with a ballistic galvanometer.

After finding the quantized character of charge, in 1891 George Stoney proposed the unit 'electron' for this fundamental unit of electrical charge. This was before the discovery of the particle by J.J. Thomson in 1897. The unit is today treated as nameless, referred to as "elementary charge", "fundamental unit of charge", or simply as "e". A measure of charge should be a multiple of the elementary charge e, even if at large scales charge seems to behave as a real quantity. In some contexts it is meaningful to speak of fractions of a charge; for example in the charging of a capacitor, or in the fractional quantum Hall effect.

In systems of units other than SI such as cgs, electric charge is expressed as combination of only three fundamental quantities (length, mass, and time), and not four, as in SI, where electric charge is a combination of length, mass, time, and electric current.

History

As reported by the ancient Greek mathematician Thales of Miletus around 600 BC, charge (or *electricity*) could be accumulated by rubbing fur on various substances, such as amber. The Greeks noted that the charged amber buttons could attract light objects such as hair. They also noted that if they rubbed the amber for long enough, they could even get an electric spark to jump. This property derives from the triboelectric effect.

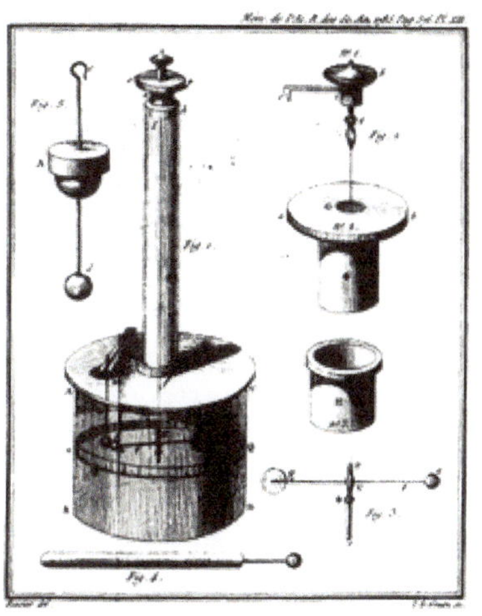

Coulomb's torsion balance

He was followed in 1660 by Otto von Guericke, who invented what was probably the first electrostatic generator. Other European pioneers were Robert Boyle, who in 1675 stated that electric attraction and repulsion can act across a vacuum; Stephen Gray, who in 1729 classified materials as conductors and insulators; and C. F. du Fay, who proposed in 1733 that electricity comes in two varieties that cancel each other, and expressed this in terms of a two-fluid theory. When glass was rubbed with silk, du Fay said that the glass was charged with *vitreous electricity*, and, when amber was rubbed with fur, the amber was charged with *resinous electricity*. In 1839, Michael Faraday showed that the apparent division between static electricity, current electricity, and bioelectricity was incorrect, and all were a consequence of the behavior of a single kind of electricity appearing in opposite polarities. It is arbitrary which polarity is called positive and which is called negative. Positive charge can be defined as the charge left on a glass rod after being rubbed with silk.

One of the foremost experts on electricity in the 18th century was Benjamin Franklin, who argued in favour of a one-fluid theory of electricity. Franklin imagined electricity as being a type of invisible fluid present in all matter; for example, he believed that it was the glass in a Leyden jar that held the accumulated charge. He posited that rubbing insulating surfaces together caused this fluid to change location, and that a flow of this fluid constitutes an electric current. He also posited that when matter contained too little of the fluid it was "negatively" charged, and when it had an excess it was "positively" charged. For a reason that was not recorded, he identified the term "positive" with vitreous electricity and "negative" with resinous electricity. William Watson arrived at the same explanation at about the same time.

Static Electricity and Electric Current

Static electricity and electric current are two separate phenomena. They both involve electric charge, and may occur simultaneously in the same object. Static electricity refers to the electric charge of an object and the related electrostatic discharge when two objects are brought together that are not at equilibrium. An electrostatic discharge creates a change in the charge of each of the two objects. In contrast, electric current is the flow of electric charge through an object, which produces no net loss or gain of electric charge.

Electrification by Friction

When a piece of glass and a piece of resin—neither of which exhibit any electrical properties—are rubbed together and left with the rubbed surfaces in contact, they still exhibit no electrical properties. When separated, they attract each other.

A second piece of glass rubbed with a second piece of resin, then separated and suspended near the former pieces of glass and resin causes these phenomena:

- The two pieces of glass repel each other.
- Each piece of glass attracts each piece of resin.
- The two pieces of resin repel each other.

This attraction and repulsion is an *electrical phenomena*, and the bodies that exhibit them are said to be *electrified*, or *electrically charged*. Bodies may be electrified in many other ways, as well as by friction. The electrical properties of the two pieces of glass are similar to each other but opposite to those of the two pieces of resin: The glass attracts what the resin repels and repels what the resin attracts.

If a body electrified in any manner whatsoever behaves as the glass does, that is, if it repels the glass and attracts the resin, the body is said to be *vitreously* electrified, and if it attracts the glass and repels the resin it is said to be *resinously* electrified. All electrified bodies are either vitreously or resinously electrified.

An established convention in the scientific community defines vitreous electrification as positive, and resinous electrification as negative. The exactly opposite properties of the two kinds of electrification justify our indicating them by opposite signs, but the application of the positive sign to one rather than to the other kind must be considered as a matter of arbitrary convention—just as it is a matter of convention in mathematical diagram to reckon positive distances towards the right hand.

No force, either of attraction or of repulsion, can be observed between an electrified body and a body not electrified.

Actually, all bodies are electrified, but may appear not electrified because of the relatively similar charge of neighboring objects in the environment. An object further electrified + or – creates an equivalent or opposite charge by default in neighboring objects, until those charges can equalize. The effects of attraction can be observed in high-voltage experiments, while lower voltage effects are merely weaker and therefore less obvious. The attraction and repulsion forces are codified by Coulomb's law (attraction falls off at the square of the distance, which has a corollary for acceleration in a gravitational field, suggesting that gravitation may be merely electrostatic phenomenon between relatively weak charges in terms of scale).

It is now known that the Franklin-Watson model was fundamentally correct. There is only one kind of electrical charge, and only one variable is required to keep track of the amount of charge. On the other hand, just knowing the charge is not a complete description of the situation. Matter is composed of several kinds of electrically charged particles, and these particles have many properties, not just charge.

The most common charge carriers are the positively charged proton and the negatively charged electron. The movement of any of these charged particles constitutes an electric current. In many situations, it suffices to speak of the *conventional current* without regard to whether it is carried by positive charges moving in the direction of the conventional current or by negative charges moving in the opposite direction. This macroscopic viewpoint is an approximation that simplifies electromagnetic concepts and calculations.

At the opposite extreme, if one looks at the microscopic situation, one sees there are many ways of carrying an electric current, including: a flow of electrons; a flow of electron "holes" that act like positive particles; and both negative and positive particles (ions or other charged particles) flowing in opposite directions in an electrolytic solution or a plasma.

Beware that, in the common and important case of metallic wires, the direction of the conventional current is opposite to the drift velocity of the actual charge carriers; i.e., the electrons. This is a source of confusion for beginners.

Conservation of Electric Charge

The total electric charge of an isolated system remains constant regardless of changes within the system itself. This law is inherent to all processes known to physics and can be derived in a local form from gauge invariance of the wave function. The conservation of charge results in the charge-current continuity equation. More generally, the net change in charge density ρ within a volume of integration V is equal to the area integral over the current density J through the closed surface $S = \partial V$, which is in turn equal to the net current I:

$$-\frac{d}{dt}\int_V \rho \mathrm{d}V = \oiint \mathbf{J}\cdot\mathrm{d}\mathbf{S} = \int J\mathrm{d}S\cos\theta = I.$$

Thus, the conservation of electric charge, as expressed by the continuity equation, gives the result:

$$I = -\frac{\mathrm{d}Q}{\mathrm{d}t}.$$

The charge transferred between times t_i and t_f is obtained by integrating both sides:

$$Q = \int_{t_\mathrm{i}}^{t_\mathrm{f}} I\,\mathrm{d}t$$

where I is the net outward current through a closed surface and Q is the electric charge contained within the volume defined by the surface.

Relativistic Invariance

Aside from the properties described in articles about electromagnetism, charge is a relativistic invariant. This means that any particle that has charge Q, no matter how fast it goes, always has charge Q. This property has been experimentally verified by showing that the charge of *one* helium nucleus (two protons and two neutrons bound together in a nucleus and moving around at high speeds) is the same as *two* deuterium nuclei (one proton and one neutron bound together, but moving much more slowly than they would if they were in a helium nucleus).

Electric Current

An electric current is a flow of electric charge. In electric circuits this charge is often carried by moving electrons in a wire. It can also be carried by ions in an electrolyte, or by both ions and electrons such as in a plasma.

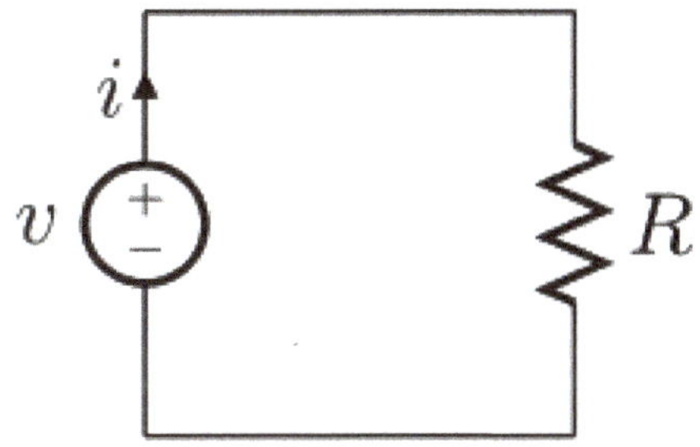

A simple electric circuit, where current is represented by the letter *i*. The relationship between the voltage (V), resistance (R), and current (I) is V=IR; this is known as Ohm's Law.

The SI unit for measuring an electric current is the ampere, which is the flow of electric charge across a surface at the rate of one coulomb per second. Electric current is measured using a device called an ammeter.

Electric currents cause Joule heating, which creates light in incandescent light bulbs. They also create magnetic fields, which are used in motors, inductors and generators.

The particles that carry the charge in an electric current are called charge carriers. In metals, one or more electrons from each atom are loosely bound to the atom, and can move freely about within the metal. These conduction electrons are the charge carriers in metal conductors.

Symbol

The conventional symbol for current is I, which originates from the French phrase *intensité de courant*, meaning *current intensity*. Current intensity is often referred to simply as *current*. The I symbol was used by André-Marie Ampère, after whom the unit of electric current is named, in formulating the eponymous Ampère's force law, which he discovered in 1820. The notation travelled from France to Great Britain, where it became standard, although at least one journal did not change from using C to I until 1896.

Conventions

In a conductive material, the moving charged particles which constitute the electric current are called charge carriers. In metals, which make up the wires and other conductors in most electrical circuits, the positively charged atomic nuclei are held in a fixed position, and the negatively charged electrons are free to move, carrying their charge from one place to another. In other materials, notably the semiconductors, the charge carriers can be positive *or* negative, depending on the dopant used. Positive and negative charge carriers may even be present at the same time, as happens in an electrochemical cell.

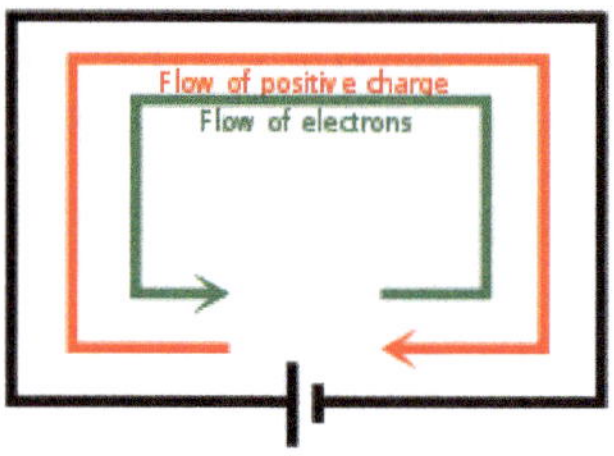

The electrons, the charge carriers in an electrical circuit, flow in the opposite direction of the conventional electric current.

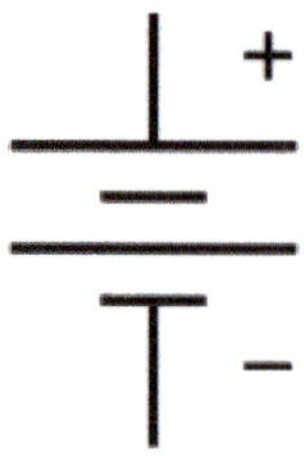

The symbol for a battery in a circuit diagram.

A flow of positive charges gives the same electric current, and has the same effect in a circuit, as an equal flow of negative charges in the opposite direction. Since current can be the flow of either positive or negative charges, or both, a convention is needed for the direction of current that is independent of the type of charge carriers. The direction of *conventional current* is arbitrarily defined as the same direction as positive charges flow.

The consequence of this convention is that electrons, the charge carriers in metal wires and most other parts of electric circuits, flow in the opposite direction of conventional current flow in an electrical circuit.

Reference Direction

Since the current in a wire or component can flow in either direction, when a variable I is defined to represent that current, the direction representing positive current must be specified, usually by an arrow on the circuit schematic diagram. This is called the *reference direction* of current I. If the current flows in the opposite direction, the variable I has a negative value.

When analyzing electrical circuits, the actual direction of current through a specific circuit element is usually unknown. Consequently, the reference directions of currents are often assigned arbitrarily. When the circuit is solved, a negative value for the variable means that the actual direction of current through that circuit element is opposite that of the chosen reference direction. In electronic circuits, the reference current directions are often chosen so that all currents are toward ground. This often corresponds to the actual current direction, because in many circuits the power supply voltage is positive with respect to ground.

Ohm's Law

Ohm's law states that the current through a conductor between two points is directly proportional to the potential difference across the two points. Introducing the constant of proportionality, the resistance, one arrives at the usual mathematical equation that describes this relationship:

where I is the current through the conductor in units of amperes, V is the potential difference measured *across* the conductor in units of volts, and R is the resistance of the conductor in units of ohms. More specifically, Ohm's law states that the R in this relation is constant, independent of the current.

AC and DC

The abbreviations *AC* and *DC* are often used to mean simply *alternating* and *direct*, as when they modify *current* or *voltage*.

Direct Current

Direct current (DC) is the unidirectional flow of electric charge. Direct current is produced by sources such as batteries, thermocouples, solar cells, and commutator-type electric machines of the dynamo type. Direct current may flow in a conductor such as a wire, but can also flow through semiconductors, insulators, or even through a vacuum as in electron or ion beams. The electric charge flows in a constant direction, distinguishing it from alternating current (AC). A term formerly used for *direct current* was galvanic current.

Alternating Current

In alternating current (AC, also ac), the movement of electric charge periodically reverses direction. In direct current (DC, also dc), the flow of electric charge is only in one direction.

AC is the form of electric power delivered to businesses and residences. The usual waveform of an AC power circuit is a sine wave. Certain applications use different waveforms, such as triangular or square waves. Audio and radio signals carried on electrical wires are also examples of alternating current. An important goal in these applications is recovery of information encoded (or *modulated*) onto the AC signal.

Occurrences

Natural observable examples of electrical current include lightning, static electricity, and the solar wind, the source of the polar auroras.

Man-made occurrences of electric current include the flow of conduction electrons in metal wires such as the overhead power lines that deliver electrical energy across long distances and the smaller wires within electrical and electronic equipment. Eddy currents are electric currents that occur in conductors exposed to changing magnetic fields. Similarly, electric currents occur, particularly in the surface, of conductors exposed to electromagnetic waves. When oscillating electric currents flow at the correct voltages within radio antennas, radio waves are generated.

In electronics, other forms of electric current include the flow of electrons through resistors or through the vacuum in a vacuum tube, the flow of ions inside a battery or a neuron, and the flow of holes within a semiconductor.

Current Measurement

Current can be measured using an ammeter.

At the circuit level, there are various techniques that can be used to measure current:

- Shunt resistors
- Hall effect current sensor transducers
- Transformers (however DC cannot be measured)
- Magnetoresistive field sensors

Resistive Heating

Joule heating, also known as *ohmic heating* and *resistive heating*, is the process by which the passage of an electric current through a conductor releases heat. It was first studied by James Prescott Joule in 1841. Joule immersed a length of wire in a fixed mass of water and measured the temperature rise due to a known current through the wire for a 30 minute period. By varying the current and the length of the wire he deduced that the heat produced was proportional to the square of the current multiplied by the electrical resistance of the wire.

$$Q \propto I^2 R$$

This relationship is known as Joule's First Law. The SI unit of energy was subsequently named the joule and given the symbol J. The commonly known unit of power, the watt, is equivalent to one joule per second.

Electromagnetism

Electromagnet

In an electromagnet a coil, of a large number of circular turns of insulated wire, wrapped on a cylindrical core, behaves like a magnet when an electric current flows through it. When the current is switched off, the coil loses its magnetism immediately. We call such a device as an electromagnet.

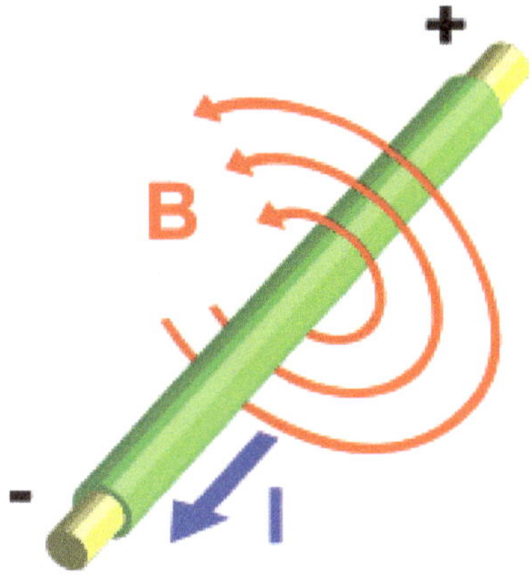

According to Ampère's law, an electric current produces a magnetic field.

Electric current produces a magnetic field. The magnetic field can be visualized as a pattern of circular field lines surrounding the wire that persists as long as there is current.

Magnetism can also produce electric currents. When a changing magnetic field is applied to a conductor, an Electromotive force (EMF) is produced, and when there is a suitable path, this causes current.

Electric current can be directly measured with a galvanometer, but this method involves breaking the electrical circuit, which is sometimes inconvenient. Current can also be measured without breaking the circuit by detecting the magnetic field associated with the current. Devices used for this include Hall effect sensors, current clamps, current transformers, and Rogowski coils.

Radio Waves

When an electric current flows in a suitably shaped conductor at radio frequencies radio waves can be generated. These travel at the speed of light and can cause electric currents in distant conductors.

Conduction Mechanisms in Various Media

In metallic solids, electric charge flows by means of electrons, from lower to higher electrical potential. In other media, any stream of charged objects (ions, for example) may constitute an electric current. To provide a definition of current independent of the type of charge carriers, *conventional current* is defined as moving in the same direction as the positive charge flow. So, in metals where the charge carriers (electrons) are negative, conventional current is in the opposite direction as the electrons. In conductors where the charge carriers are positive, conventional current is in the same direction as the charge carriers.

In a vacuum, a beam of ions or electrons may be formed. In other conductive materials, the electric current is due to the flow of both positively and negatively charged particles at the same time. In still others, the current is entirely due to positive charge flow. For example, the electric currents in electrolytes are flows of positively and negatively charged ions. In a common lead-acid electrochemical cell, electric currents are composed of positive hydrogen ions (protons) flowing in one direction, and negative sulfate ions flowing in the other. Electric currents in sparks or plasma are flows of electrons as well as positive and negative ions. In ice and in certain solid electrolytes, the electric current is entirely composed of flowing ions.

Metals

In a metal, some of the outer electrons in each atom are not bound to the individual atom as they are in insulating materials, but are free to move within the metal lattice. These conduction electrons can serve as charge carriers, carrying a current. Metals are particularly conductive because there are a large number of these free electrons, typically one per atom in the lattice. With no external electric field applied, these electrons move about randomly due to thermal energy but, on average, there is zero net current within the metal. At room temperature, the average speed of these random motions is 10^6 metres per second. Given a surface through which a metal wire passes, electrons move in both directions across the surface at an equal rate. As George Gamow wrote in his popular science book, *One, Two, Three...Infinity* (1947), "The metallic substances differ from all other materials by the fact that the outer shells of their atoms are bound rather loosely, and often let one of their electrons go free. Thus the interior of a metal is filled up with a large number of unattached electrons that travel aimlessly around like a crowd of displaced persons. When a metal wire is subjected to electric force applied on its opposite ends, these free electrons rush in the direction of the force, thus forming what we call an electric current."

When a metal wire is connected across the two terminals of a DC voltage source such as a battery, the source places an electric field across the conductor. The moment contact is made, the free electrons of the conductor are forced to drift toward the positive terminal under the influence of this field. The free electrons are therefore the charge carrier in a typical solid conductor.

For a steady flow of charge through a surface, the current I (in amperes) can be calculated with the following equation:

$$I = \frac{Q}{t},$$

where Q is the electric charge transferred through the surface over a time t. If Q and t are measured in coulombs and seconds respectively, I is in amperes.

More generally, electric current can be represented as the rate at which charge flows through a given surface as:

$$I = \frac{dQ}{dt}.$$

Electrolytes

Electric currents in electrolytes are flows of electrically charged particles (ions). For example, if an electric field is placed across a solution of Na^+ and Cl^- (and conditions are right) the sodium ions move towards the negative electrode (cathode), while the chloride ions move towards the positive electrode (anode). Reactions take place at both electrode surfaces, absorbing each ion.

Water-ice and certain solid electrolytes called proton conductors contain positive hydrogen ions ("protons") that are mobile. In these materials, electric currents are composed of moving protons, as opposed to the moving electrons in metals.

In certain electrolyte mixtures, brightly coloured ions are the moving electric charges. The slow progress of the colour makes the current visible.

Gases And Plasmas

In air and other ordinary gases below the breakdown field, the dominant source of electrical conduction is via relatively few mobile ions produced by radioactive gases, ultraviolet light, or cosmic rays. Since the electrical conductivity is low, gases are dielectrics or insulators. However, once the applied electric field approaches the breakdown value, free electrons become sufficiently accelerated by the electric field to create additional free electrons by colliding, and ionizing, neutral gas atoms or molecules in a process called avalanche breakdown. The breakdown process forms a plasma that contains enough mobile electrons and positive ions to make it an electrical conductor. In the process, it forms a light emitting conductive path, such as a spark, arc or lightning.

Plasma is the state of matter where some of the electrons in a gas are stripped or "ionized" from their molecules or atoms. A plasma can be formed by high temperature, or by application of a high electric or alternating magnetic field as noted above. Due to their lower mass, the electrons in a plasma accelerate more quickly in response to an electric field than the heavier positive ions, and hence carry the bulk of the current. The free ions recombine to create new chemical compounds (for example, breaking atmospheric oxygen into single oxygen [$O_2 \rightarrow 2O$], which then recombine creating ozone [O_3]).

Vacuum

Since a "perfect vacuum" contains no charged particles, it normally behaves as a perfect insulator. However, metal electrode surfaces can cause a region of the vacuum to become conductive by injecting free electrons or ions through either field electron emission or thermionic emission. Thermionic emission occurs when the thermal energy exceeds the metal's work function, while field electron emission occurs when the electric field at the surface of the metal is high enough to cause tunneling, which results in the ejection of free electrons from the metal into the vacuum. Externally heated electrodes are often used to generate an electron cloud as in the filament or indirectly heated cathode of vacuum tubes. Cold electrodes can also spontaneously produce electron clouds via thermionic emission when small incandescent regions (called cathode spots or anode spots) are formed. These are incandescent regions of the electrode surface that are created by a localized high current. These regions may be initiated by field electron emission, but are then sustained by localized thermionic emission once a vacuum arc forms. These small electron-emitting regions can form quite rapidly, even explosively, on a metal surface subjected to a high electrical field. Vacuum tubes and sprytrons are some of the electronic switching and amplifying devices based on vacuum conductivity.

Superconductivity

Superconductivity is a phenomenon of exactly zero electrical resistance and expulsion of magnetic fields occurring in certain materials when cooled below a characteristic critical temperature. It was discovered by Heike Kamerlingh Onnes on April 8, 1911 in Leiden. Like ferromagnetism and atomic spectral lines, superconductivity is a quantum mechanical phenomenon. It is characterized by the Meissner effect, the complete ejection of magnetic field lines from the interior of the superconductor as it transitions into the superconducting state. The occurrence of the Meissner effect indicates that superconductivity cannot be understood simply as the idealization of *perfect conductivity* in classical physics.

Semiconductor

In a semiconductor it is sometimes useful to think of the current as due to the flow of positive "holes" (the mobile positive charge carriers that are places where the semiconductor crystal is missing a valence electron). This is the case in a p-type semiconductor. A semiconductor has electrical conductivity intermediate in magnitude between that of a conductor and an insulator. This means a conductivity roughly in the range of 10^{-2} to 10^4 siemens per centimeter (S⍰cm^{-1}).

In the classic crystalline semiconductors, electrons can have energies only within certain bands (i.e. ranges of levels of energy). Energetically, these bands are located between the energy of the ground state, the state in which electrons are tightly bound to the atomic nuclei of the material, and the free electron energy, the latter describing the energy required for an electron to escape entirely from the material. The energy bands each correspond to a large number of discrete quantum states of the electrons, and most of the states with low energy (closer to the nucleus) are occupied, up to a particular band called the *valence band*. Semiconductors and insulators are distinguished from metals because the valence band in any given metal is nearly filled with electrons under usual operating conditions, while very few (semiconductor) or virtually none (insulator) of them are available in the *conduction band*, the band immediately above the valence band.

The ease of exciting electrons in the semiconductor from the valence band to the conduction band depends on the band gap between the bands. The size of this energy band gap serves as an arbitrary dividing line (roughly 4 eV) between semiconductors and insulators.

With covalent bonds, an electron moves by hopping to a neighboring bond. The Pauli exclusion principle requires that the electron be lifted into the higher anti-bonding state of that bond. For delocalized states, for example in one dimension – that is in a nanowire, for every energy there is a state with electrons flowing in one direction and another state with the electrons flowing in the other. For a net current to flow, more states for one direction than for the other direction must be occupied. For this to occur, energy is required, as in the semiconductor the next higher states lie above the band gap. Often this is stated as: full bands do not contribute to the electrical conductivity. However, as a semiconductor's temperature rises above absolute zero, there is more energy in the semiconductor to spend on lattice vibration and on exciting electrons into the conduction band. The current-carrying electrons in the conduction band are known as *free electrons*, though they are often simply called *electrons* if that is clear in context.

Current Density and Ohm's Law

Current density is a measure of the density of an electric current. It is defined as a vector whose magnitude is the electric current per cross-sectional area. In SI units, the current density is measured in amperes per square metre.

$$I = \int \vec{J} \cdot d\vec{A}$$

where I is current in the conductor, $\vec{J}$ is the current density, and $d\vec{A}$ is the differential cross-sectional area vector.

The current density (current per unit area) $\vec{J}$ in materials with finite resistance is directly proportional to the electric field $\vec{E}$ in the medium. The proportionality constant is called the conductivity σ of the material, whose value depends on the material concerned and, in general, is dependent on the temperature of the material:

$$\vec{J} = \sigma \vec{E}$$

The reciprocal of the conductivity σ of the material is called the resistivity ρ of the material and the above equation, when written in terms of resistivity becomes:

$$\vec{J} = \frac{\vec{E}}{\rho} \text{ or}$$

$$\vec{E} = \rho \vec{J}$$

Conduction in semiconductor devices may occur by a combination of drift and diffusion, which is proportional to diffusion constant D and charge density α_q. The current density is then:

$$J = \sigma E + Dq\nabla n,$$

with q being the elementary charge and n the electron density. The carriers move in the direction of decreasing concentration, so for electrons a positive current results for a positive density gradient. If the carriers are holes, replace electron density n by the negative of the hole density p.

In linear anisotropic materials, σ, ρ and D are tensors.

In linear materials such as metals, and under low frequencies, the current density across the conductor surface is uniform. In such conditions, Ohm's law states that the current is directly proportional to the potential difference between two ends (across) of that metal (ideal) resistor (or other ohmic device):

$$I = \frac{V}{R},$$

where I is the current, measured in amperes; V is the potential difference, measured in volts; and R is the resistance, measured in ohms. For alternating currents, especially at higher frequencies, skin effect causes the current to spread unevenly across the conductor cross-section, with higher density near the surface, thus increasing the apparent resistance.

Drift Speed

The mobile charged particles within a conductor move constantly in random directions, like the particles of a gas. To create a net flow of charge, the particles must also move together with an average drift rate. Electrons are the charge carriers in metals and they follow an erratic path, bouncing from atom to atom, but generally drifting in the opposite direction of the electric field. The speed they drift at can be calculated from the equation:

$$I = nAvQ,$$

where

I is the electric current

n is number of charged particles per unit volume (or charge carrier density)

A is the cross-sectional area of the conductor

v is the drift velocity, and

Q is the charge on each particle.

Typically, electric charges in solids flow slowly. For example, in a copper wire of cross-section 0.5 mm^2, carrying a current of 5 A, the drift velocity of the electrons is on the order of a millimetre per second. To take a different example, in the near-vacuum inside a cathode ray tube, the electrons travel in near-straight lines at about a tenth of the speed of light.

Any accelerating electric charge, and therefore any changing electric current, gives rise to an electromagnetic wave that propagates at very high speed outside the surface of the conductor. This speed is usually a significant fraction of the speed of light, as can be deduced from Maxwell's Equations, and is therefore many times faster than the drift velocity of the electrons. For example, in AC power lines, the waves of electromagnetic energy propagate through the space between the wires, moving from a source to a distant load, even though the electrons in the wires only move back and forth over a tiny distance.

The ratio of the speed of the electromagnetic wave to the speed of light in free space is called the velocity factor, and depends on the electromagnetic properties of the conductor and the insulating materials surrounding it, and on their shape and size.

The magnitudes (but, not the natures) of these three velocities can be illustrated by an analogy with the three similar velocities associated with gases.

- The low drift velocity of charge carriers is analogous to air motion; in other words, winds.
- The high speed of electromagnetic waves is roughly analogous to the speed of sound in a gas (these waves move through the medium much faster than any individual particles do)
- The random motion of charges is analogous to heat – the thermal velocity of randomly vibrating gas particles.

Cathode

A cathode is the electrode from which a conventional current leaves a polarized electrical device. (This definition can be recalled by using the mnemonic *CCD* for *cathode current departs.*) A conventional current describes the direction in which positive electronic charges move. Electrons have a negative charge, so the movement of electrons is opposite to the conventional current flow. Consequently, the mnemonic *cathode current departs* also means that electrons flow into the device's cathode.

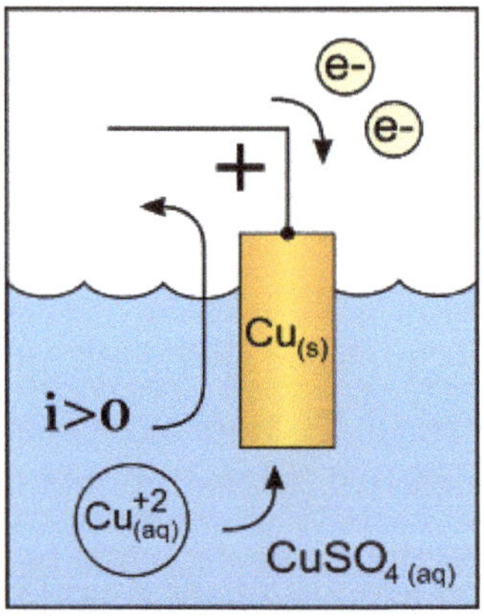

Diagram of a copper cathode in a galvanic cell (e.g., a battery). A positive current *i* flows out of the cathode.

Cathode polarity with respect to the anode can be positive or negative; it depends on how the device operates. Although positively charged cations always move towards the cathode (hence their name) and negatively charged anions move away from it, cathode polarity depends on the device type, and can even vary according to the operating mode. In a device which takes energy (such as recharging a battery), the cathode is negative, and in a device which provides energy (such as discharging a battery), the cathode is positive:

- In a discharging battery or a galvanic cell, the cathode is the positive terminal since that is where the current flows out of the device. This outward current is carried internally by positive ions moving from the electrolyte to the positive cathode (chemical energy is responsible for this "uphill" motion). It is continued externally by electrons moving inwards, this negative charge moving inwards constituting positive current flowing outwards. For example, the Daniell galvanic cell's copper electrode is the positive terminal and the cathode.
- In a recharging battery, or an electrolytic cell performing electrolysis, the cathode is the negative terminal, from which current exits the device and returns to the external generator. For example, reversing the current direction in a Daniell galvanic cell would produce an electrolytic cell, where the copper electrode is the positive terminal and the anode.

- In a diode, the cathode is the negative terminal at the pointed end of the arrow symbol, where current flows out of the device. Note: electrode naming for diodes is always based on the direction of the forward current (that of the arrow, in which the current flows "most easily"), even for types such as Zener diodes or solar cells where the current of interest is the reverse current.
- In vacuum tubes (including cathode ray tubes) it is the negative terminal where electrons enter the device from the external circuit and proceed into the tube's near vacuum, constituting a positive current flowing out of the device.

An electrode through which current flows the other way (into the device) is termed an anode.

Etymology

The word was coined in 1834, 'descent' or 'way down', by William Whewell, who had been consulted by Michael Faraday over some new names needed to complete a paper on the recently discovered process of electrolysis. In that paper Faraday explained that when an electrolytic cell is oriented so that electric current traverses the "decomposing body" (electrolyte) in a direction "from East to West, or, which will strengthen this help to the memory, that in which the sun appears to move", the cathode is where the current leaves the electrolyte, on the West side: "*kata* downwards, `*odos* a way ; the way which the sun sets".

The use of 'West' to mean the 'out' direction (actually 'out' → 'West' → 'sunset' → 'down', i.e. 'out of view') may appear unnecessarily contrived. Previously, as related in the first reference cited above, Faraday had used the more straightforward term "exode" (the doorway where the current exits). His motivation for changing it to something meaning 'the West electrode' (other candidates had been "westode", "occiode" and "dysiode") was to make it immune to a possible later change in the direction convention for current, whose exact nature was not known at the time. The reference he used to this effect was the Earth's magnetic field direction, which at that time was believed to be invariant. He fundamentally defined his arbitrary orientation for the cell as being that in which the internal current would run parallel to and in the same direction as a hypothetical magnetizing current loop around the local line of latitude which would induce a magnetic dipole field oriented like the Earth's. This made the internal current East to West as previously mentioned, but in the event of a later convention change it would have become West to East, so that the West electrode would not have been the 'way out' any more. Therefore, "exode" would have become inappropriate, whereas "cathode" meaning 'West electrode' would have remained correct with respect to the unchanged direction of the actual phenomenon underlying the current, then unknown but, he thought, unambiguously defined by the magnetic reference. In retrospect the name change was unfortunate, not only because the Greek roots alone do not reveal the cathode's function any more, but more importantly because, as we now know, the

Earth's magnetic field direction on which the "cathode" term is based is subject to reversals whereas the current direction convention on which the "exode" term was based has no reason to change in the future.

Since the later discovery of the electron, an easier to remember, and more durably technically correct (although historically false), etymology has been suggested: cathode, from the Greek *kathodos*, 'way down', 'the way (down) into the cell (or other device) for electrons'.

Flow of Electrons

The flow of electrons is almost always from anode to cathode outside of the cell or device, regardless of the cell or device type and operating mode. An exception is when a diode reverse-conducts, either by accident (breakdown of a normal diode) or by design (breakdown of a Zener diode, photo-current of a photodiode).

In Chemistry

In chemistry, a cathode is the electrode of an electrochemical cell at which reduction occurs; a useful mnemonic to remember this is AnOx RedCat (Oxidation at the Anode = Reduction at the Cathode). Another mnemonic is to note the cathode has a 'c', as does 'reduction'. Hence, reduction at the cathode. Perhaps most useful would be to remember cathode corresponds to cation (acceptor) and anode corresponds to anion (acceptor). The cathode can be negative like when the cell is electrolytic (where electrical energy provided to the cell is being used for decomposing chemical compounds); or positive like when the cell is galvanic (where chemical reactions are used for generating electrical energy). The cathode supplies electrons to the positively charged cations which flow to it from the electrolyte (even if the cell is galvanic, i.e., when the cathode is positive and therefore would be expected to repel the positively charged cations; this is due to electrode potential relative to the electrolyte solution being different for the anode and cathode metal/electrolyte systems in a galvanic cell).

The cathodic current, in electrochemistry, is the flow of electrons from the cathode interface to a species in solution. The anodic current is the flow of electrons into the anode from a species in solution.

Electrolytic Cell

In an electrolytic cell, the cathode is where the negative polarity is applied to drive the cell. Common results of reduction at the cathode are hydrogen gas or pure metal from metal ions. When discussing the relative reducing power of two redox agents, the couple for generating the more reducing species is said to be more "cathodic" with respect to the more easily reduced reagent.

Galvanic Cell

In a galvanic cell, the cathode is where the positive pole is connected to allow the circuit to be completed: as the anode of the galvanic cell gives off electrons, they return from the circuit into the cell through the cathode.

Electroplating Metal Cathode (Electrolysis)

When metal ions are reduced from ionic solution, they form a pure metal surface on the cathode. Items to be plated with pure metal are attached to and become part of the cathode in the electrolytic solution.

In Electronics

In physics or electronics, a cathode is an electrode that emits electrons into the device. This contrasts with an anode, which accepts electrons.

Vacuum Tubes

Glow from the directly heated cathode of a 1 kW power tetrode tube in a radio transmitter. The cathode filament is not directly visible

In a vacuum tube or electronic vacuum system, the cathode is a metal surface which emits free electrons into the evacuated space. Since the electrons are attracted to the positive nuclei of the metal atoms, they normally stay inside the metal and require energy to leave it; this is called the *work function* of the metal. Cathodes are induced to emit electrons by several mechanisms:

- *Thermionic emission*: The cathode can be heated. The increased thermal motion of the metal atoms "knocks" electrons out of the surface, an effect called thermionic emission. This technique is used in most vacuum tubes.
- *Field electron emission*: A strong electric field can be applied to the surface by placing an electrode with a high positive voltage near the cathode. The positively charged electrode attracts the electrons, causing some electrons to leave the cathode's surface. This process is used in cold cathodes in some electron microscopes, and in microelectronics fabrication,

- *Secondary emission*: An electron, atom or molecule colliding with the surface of the cathode with enough energy can knock electrons out of the surface. These electrons are called *secondary electrons*. This mechanism is used in gas-discharge lamps such as neon lamps.
- *Photoelectric emission*: Electrons can also be emitted from the electrodes of certain metals when light of frequency greater than the threshold frequency falls on it. This effect is called photoelectric emission, and the electrons produced are called *photoelectrons*. This effect is used in phototubes and image intensifier tubes.

Cathodes can be divided into two types:

Hot Cathode

A hot cathode is a cathode that is heated by a filament to produce electrons by thermionic emission. The filament is a thin wire of a refractory metal like tungsten heated red-hot by an electric current passing through it. Before the advent of transistors in the 1960s, virtually all electronic equipment used hot-cathode vacuum tubes. Today hot cathodes are used in vacuum tubes in radio transmitters and microwave ovens, to produce the electron beams in older cathode ray tube (CRT) type televisions and computer monitors, in x-ray machines, electron microscopes, and fluorescent tubes.

Two indirectly-heated cathodes (orange heater strip) in ECC83 dual triode tube

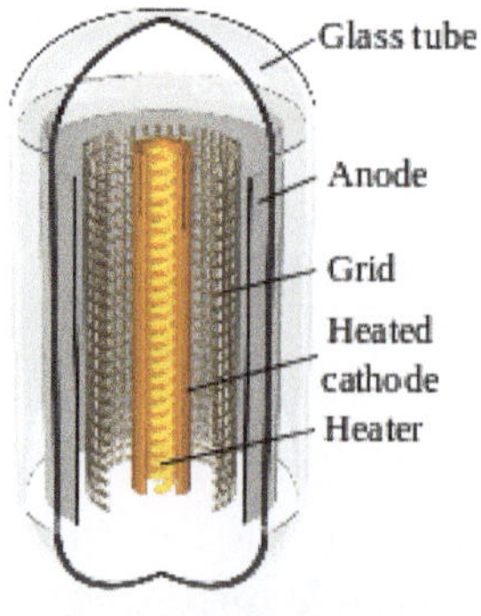

Cutaway view of a triode vacuum tube with an indirectly-heated cathode *(orange tube)*, showing the heater element inside

There are two types of hot cathodes:

- Directly heated cathode: In this type, the filament itself is the cathode and emits the electrons directly. Directly heated cathodes were used in the first vacuum tubes, but today they are only used in fluorescent tubes, some large transmitting vacuum tubes, and all X-ray tubes.

- Indirectly heated cathode: In this type, the filament is not the cathode but rather heats the cathode which then emits electrons. Indirectly heated cathodes are used in most devices today. For example, in most vacuum tubes the cathode is a nickel tube with the filament inside it, and the heat from the filament causes the outside surface of the tube to emit electrons. The filament of an indirectly heated cathode is usually called the *heater*. The main reason for using an indirectly heated cathode is to isolate the rest of the vacuum tube from the electric potential across the filament. Many vacuum tubes use alternating current to heat the filament. In a tube in which the filament itself was the cathode, the alternating electric field from the filament surface would affect the movement of the electrons and introduce hum into the tube output. It also allows the filaments in all the tubes in an electronic device to be tied together and supplied from the same current source, even though the cathodes they heat may be at different potentials.

In order to improve electron emission, cathodes are treated with chemicals, usually compounds of metals with a low work function. Treated cathodes require less surface area, lower temperatures and less power to supply the same cathode current. The untreated tungsten filaments used in early tubes (called "bright emitters") had to be heated to 2500 °F (1400 °C), white-hot, to produce sufficient thermionic emission for use, while modern coated cathodes produce far more electrons at a given temperature so they only have to be heated to 800–1100 °F (425–600 °C) There are two main types of treated cathodes:

Cold cathode *(lefthand electrode)* in neon lamp

- Coated cathode – In these the cathode is covered with a coating of alkali metal oxides, often barium and strontium oxide. These are used in low-power tubes.
- Thoriated tungsten – In high-power tubes, ion bombardment can destroy the coating on a coated cathode. In these tubes a directly heated cathode consisting of a filament made of tungsten incorporating a small amount of thorium is used. The layer of thorium on the surface which reduces the work function of the cathode is continually replenished as it is lost by diffusion of thorium from the interior of the metal.

Cold Cathode

This is a cathode that is not heated by a filament. They may emit electrons by field electron emission, and in gas-filled tubes by secondary emission. Some examples are electrodes in neon lights, cold-cathode fluorescent lamps (CCFLs) used as backlights in laptops, thyratron tubes, and Crookes tubes. They do not necessarily operate at room temperature; in some devices the cathode is heated by the electron current flowing through it to a temperature at which thermionic emission occurs. For example, in some fluorescent tubes a momentary high voltage is applied to the electrodes to start the current through the tube; after starting the electrodes are heated enough by the current to keep emitting electrons to sustain the discharge.

Cold cathodes may also emit electrons by photoelectric emission. These are often called *photocathodes* and are used in phototubes used in scientific instruments and image intensifier tubes used in night vision goggles.

Diodes

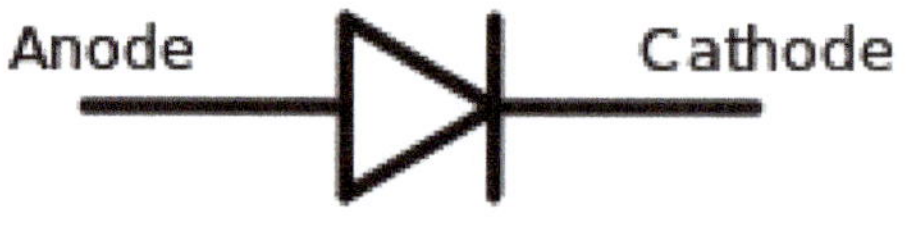

In a semiconductor diode, the cathode is the N–doped layer of the PN junction with a high density of free electrons due to doping, and an equal density of fixed positive charges, which are the dopants that have been thermally ionized. In the anode, the converse applies: It features a high density of free "holes" and consequently fixed negative dopants which have captured an electron (hence the origin of the holes).

When P and N-doped layers are created adjacent to each other, diffusion ensures that electrons flow from high to low density areas: That is, from the N to the P side. They leave behind the fixed positively charged dopants near the junction. Similarly, holes diffuse from P to N leaving behind fixed negative ionised dopants near the junction. These layers of fixed positive and negative charges are collectively known as the depletion layer because they are depleted of free electrons and holes. The depletion layer at the junction is at the origin of the diode's rectifying properties. This is due to the resulting internal field and corresponding potential barrier which inhibit current flow in reverse applied bias which increases the internal depletion layer field. Conversely, they allow it in forwards applied bias where the applied bias reduces the built in potential barrier.

Electrons which diffuse from the cathode into the P-doped layer, or anode, become what are termed "minority carriers" and tend to recombine there with the majority carriers, which are holes, on a timescale characteristic of the material which is the p-type minority carrier lifetime. Similarly, holes diffusing into the N-doped layer become minority carriers and tend to recombine with electrons. In equilibrium, with no applied bias, thermally assisted diffusion of electrons and holes in opposite directions across the depletion layer ensure a zero net current with electrons flowing from cathode to anode and recombining, and holes flowing from anode to cathode across the junction or depletion layer and recombining.

Like a typical diode, there is a fixed anode and cathode in a Zener diode, but it will conduct current in the reverse direction (electrons flow from anode to cathode) if its breakdown voltage or "Zener voltage" is exceeded.

Anode

An anode is an electrode through which conventional current flows into a polarized electrical device. A common mnemonic is ACID for "anode current into device". The direction of (positive) electric current is opposite to the direction of electron flow: (negatively charged) electrons flow out the anode to the outside circuit.

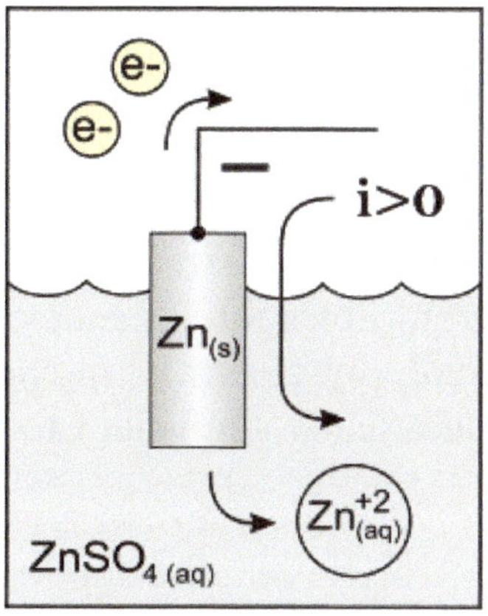

Diagram of a zinc anode in a galvanic cell. Note how electrons move out of the cell, and the conventional current moves into it in the opposite direction.

Charge Flow

The terms anode and cathode do not relate to the voltage polarity of those electrodes but the direction of the current: whether positive charge is flowing into or out of the device. Conventional current quantifies the flow of positive charge. In most cases, positive charge leaves the device via the cathode, and positive charge flows into the device via the anode.

Conventional current depends not only on the direction the charge carriers move, but also the carriers' charge. The currents outside the device are usually carried by electrons in a metal conductor. The flow of electrons is opposite to conventional current because electrons have a negative charge. Consequently, electrons leave the device via the anode, and electrons enter the device through the cathode.

The anode and cathode have slightly different definitions for electrical devices such as diodes and vacuum tubes where the electrode naming is fixed and does not depend on the actual charge flow (current). These devices usually allow substantial current flow in one direction but negligible current in the other direction. Consequently, the electrode names use the terms that have substantial ordinary currents. An ideal diode allows current in one direction but not in the other. The electrode that allows positive charges to flow into it would be the anode, but that electrode would never allow positive charges to flow out, so that ideal diode terminal could never be the cathode. For the ideal diode, it makes sense to always call that terminal the anode. For non-ideal diodes, the electrodes are also given fixed names even though such a diode under reverse bias would have a positive charge flow out of the "anode". For some operating conditions and devices, such as diode breakdown, Zener diodes, or photodiodes, the positive charge flow out of the "anode" could be substantial.

Examples

The polarity of voltage on an anode with respect to an associated cathode varies depending on the device type and on its operating mode. In the following examples, the anode is negative in a device that provides power, and positive in a device that consumes power:

In a discharging battery or galvanic cell (diagram at right), the anode is the negative terminal because it is where current flows into "the device" (i.e. the battery cell). This inward current is carried externally by electrons moving outwards, negative charge flowing in one direction being electrically equivalent to positive charge flowing in the opposite direction.

In a recharging battery, or an electrolytic cell, the anode is the positive terminal, which receives current from an external generator. The current through a recharging battery is opposite to the direction of current during discharge; in other words, the electrode which was the cathode during battery discharge becomes the anode while the battery is recharging.

In a diode, the anode is the positive terminal at the tail of the arrow symbol (flat side of the triangle), where current flows into the device. Note electrode naming for diodes is always based on the direction of the forward current (that of the arrow, in which the current flows "most easily"), even for types such as Zener diodes or solar cells where the current of interest is the reverse current.

In a cathode ray tube, the anode is the positive terminal where electrons flow out of the device, i.e., where positive electric current flows in.

Etymology

The word was coined in 1834, 'ascent', by William Whewell, who had been consulted by Michael Faraday over some new names needed to complete a paper on the recently discovered process of electrolysis. In that paper Faraday explained that when an electrolytic cell is oriented so that electric current traverses the "decomposing body" (electrolyte) in a direction "from East to West, or, which will strengthen this help to the memory, that in which the sun appears to move", the anode is where the current enters the electrolyte, on the East side: "*ano* upwards, *odos* a way; the way which the sun rises".

The use of 'East' to mean the 'in' direction (actually 'in' → 'East' → 'sunrise' → 'up') may appear contrived. Previously, as related in the first reference cited above, Faraday had used the more straightforward term "eisode" (the doorway where the current enters). His motivation for changing it to something meaning 'the East electrode' (other candidates had been "eastode", "oriode" and "anatolode") was to make it immune to a possible later change in the direction convention for current, whose exact nature was not known at the time. The reference he used to this effect was the Earth's magnetic field direction, which at that time was believed to be invariant. He fundamentally defined his arbitrary orientation for the cell as being that in which the internal current would run parallel to and in the same direction as a hypothetical magnetizing current loop around the local line of latitude which would induce a magnetic dipole field oriented like the Earth's. This made the internal current East to West as previously mentioned, but in the event of a later convention change it would have become West to East, so that the East electrode would not have been the 'way in' any more. Therefore, "eisode" would have become inappropriate, whereas "anode" meaning 'East electrode' would have remained correct with respect to the unchanged direction of the actual phenomenon underlying the current, then unknown but, he thought, unambiguously defined by the magnetic reference. In retrospect the name change was unfortunate, not only because the Greek roots alone do not reveal the anode's function any more, but more importantly because as we now know, the Earth's magnetic field direction on which the "anode" term is based is subject to reversals whereas the current direction convention on which the "eisode" term was based has no reason to change in the future.

Since the later discovery of the electron, an easier to remember and more durably correct technically although historically false, etymology has been suggested: anode, from the Greek *anodos*, 'way up', 'the way (up) out of the cell (or other device) for electrons'.

Electrolytic Anode

In electrochemistry, the *anode* is where oxidation occurs and is the positive polarity contact in an electrolytic cell. At the anode, anions (negative ions) are forced by the electrical potential to react chemically and give off electrons (oxidation) which then flow up and into the driving circuit. Mnemonics: LEO Red Cat (Loss of Electrons is Oxidation, Reduction occurs at the Cathode), or AnOx Red Cat (Anode Oxidation, Reduction Cathode), or OIL RIG (Oxidation is Loss, Reduction is Gain of electrons), or Roman Catholic and Orthodox (Reduction – Cathode, anode – Oxidation), or LEO the lion says GER (Losing electrons is Oxidation, Gaining electrons is Reduction).

This process is widely used in metals refining. For example, in copper refining, copper anodes, an intermediate product from the furnaces, are electrolysed in an appropriate solution (such as sulfuric acid) to yield high purity (99.99%) cathodes. Copper cathodes produced using this method are also described as electrolytic copper.

Battery or Galvanic Cell Anode

In a battery or galvanic cell, the anode is the negative electrode from which electrons flow out towards the external part of the circuit. Internally the positively charged cations are flowing away from the anode (even though it is negative and therefore would be expected to attract them, this is due to electrode potential relative to the electrolyte solution being different for the anode and cathode metal/electrolyte systems); but, external to the cell in the circuit, electrons are being pushed out through the negative contact and thus through the circuit by the voltage potential as would be expected. Note: in a galvanic cell, contrary to what occurs in an electrolytic cell, no anions flow to the anode, the internal current being entirely accounted for by the cations flowing away from it (cf drawing).

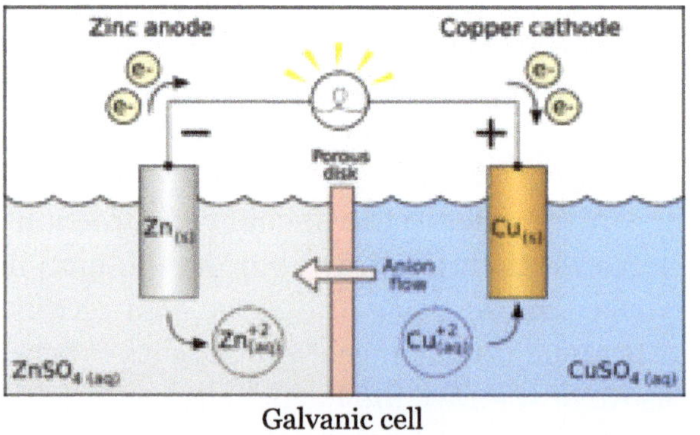

Galvanic cell

In the United States, many battery manufacturers regard the positive electrode as the anode, particularly in their technical literature. Though technically incorrect, it does resolve the problem of which electrode is the anode in a secondary (or rechargeable) cell. Using the traditional definition, the anode switches ends between charge and discharge cycles.

Vacuum Tube Anode

In electronic vacuum devices such as a cathode ray tube, the anode is the positively charged electron collector. In a tube, the anode is a charged positive plate that collects the electrons emitted by the cathode through electric attraction. It also accelerates the flow of these electrons.

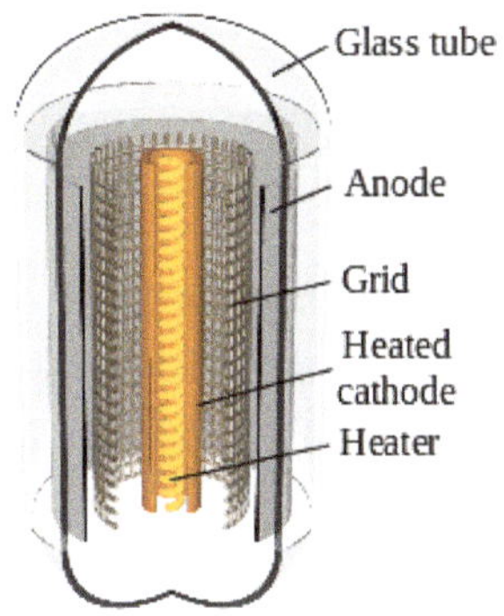

Cutaway diagram of a triode vacuum tube, showing the plate (anode)

Diode Anode

In a semiconductor diode, the anode is the P-doped layer which initially supplies *holes* to the junction. In the junction region, the holes supplied by the anode combine with electrons supplied from the N-doped region, creating a depleted zone. As the P-doped layer supplies holes to the depleted region, negative dopant ions are left behind in the P-doped layer ('P' for positive charge-carrier ions). This creates a base negative charge on the anode. When a positive voltage is applied to anode of the diode from the circuit, more *holes* are able to be transferred to the depleted region, and this causes the diode to become conductive, allowing current to flow through the circuit. The terms anode and cathode should not be applied to a Zener diode, since it allows flow in either direction, depending on the polarity of the applied potential (i.e. voltage).

Sacrificial Anode

In cathodic protection, a metal anode that is more reactive to the corrosive environment of the system to be protected is electrically linked to the protected system, and partially corrodes or dissolves, which protects the metal of the system it is connected to. As an example, an iron or steel ship's hull may be protected by a zinc sacrificial anode, which will dissolve into the seawater and prevent the hull from being corroded. Sacrificial anodes are particularly needed for systems where a static charge is generated by the action of flowing liquids, such as pipelines and watercraft. Sacrificial anodes are also generally used in tank-type water heaters.

Sacrificial anodes mounted “on the fly” for corrosion protection of a metal structure

In 1824 to reduce the impact of this destructive electrolytic action on ships hulls, their fastenings and underwater equipment, the scientist-engineer Sir Humphry Davy, developed the first and still most widely used marine electrolysis protection system. Davy installed sacrificial anodes made from a more electrically reactive (less noble) metal attached to the vessel hull and electrically connected to form a cathodic protection circuit.

A less obvious example of this type of protection is the process of galvanising iron. This process coats iron structures (such as fencing) with a coating of zinc metal. As long as the zinc remains intact, the iron is protected from the effects of corrosion. Inevitably, the zinc coating becomes breached, either by cracking or physical damage. Once this occurs, corrosive elements act as an electrolyte and the zinc/iron combination as electrodes. The resultant current ensures that the zinc coating is sacrificed but that the base iron does not corrode. Such a coating can protect an iron structure for a few decades, but once the protecting coating is consumed, the iron rapidly corrodes.

If, conversely, tin is used to coat steel, when a breach of the coating occurs it actually accelerates oxidation of the iron.

Related Antonym

The opposite of an anode is a cathode. When the current through the device is reversed, the electrodes switch functions, so anode becomes cathode, while cathode becomes anode, as long as the reversed current is applied, with the exception of diodes where electrode naming is always based on the forward current direction.

Chemical Kinetics

Chemical kinetics, also known as reaction kinetics, is the study of rates of chemical processes. Chemical kinetics includes investigations of how different experimental conditions can influence the speed of a chemical reaction and yield information about the reaction's mechanism and transition states, as well as the construction of mathematical models that can describe the characteristics of a chemical reaction.

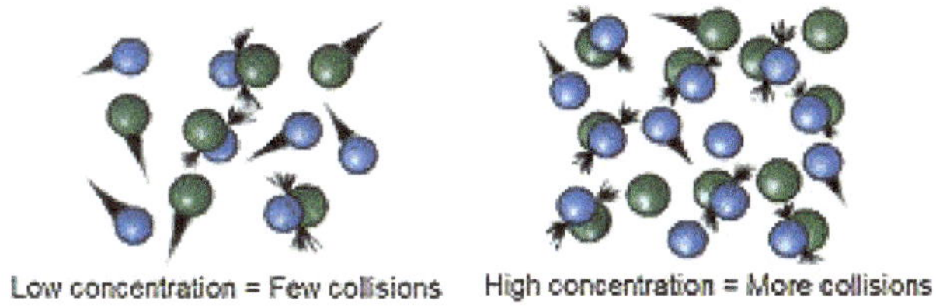

Reaction rate tends to increase with concentration – a phenomenon explained by collision theory.

History

In 1864, Peter Waage and Cato Guldberg pioneered the development of chemical kinetics by formulating the law of mass action, which states that the speed of a chemical reaction is proportional to the quantity of the reacting substances.

Van 't Hoff studied chemical dynamics and published in 1884 his famous "Etudes de dynamique chimique". In 1901 he was awarded by the first Nobel Prize in Chemistry "in recognition of the extraordinary services he has rendered by the discovery of the laws of chemical dynamics and osmotic pressure in solutions". After van 't Hoff, chemical kinetics deals with the experimental determination of reaction rates from which rate laws and rate constants are derived. Relatively simple rate laws exist for zero order reactions (for which reaction rates are independent of concentration), first order reactions, and second order reactions, and can be derived for others. Elementary reactions follow the law of mass action, but the rate law of stepwise reactions has to be derived by combining the rate laws of the various elementary steps, and can become rather complex. In consecutive reactions, the rate-determining step often determines the kinetics. In consecutive first order reactions, a steady state approximation can simplify the rate law. The activation energy for a reaction is experimentally determined through the Arrhenius equation and the Eyring equation. The main factors that influence the reaction rate include: the physical state of the reactants, the concentrations of the reactants, the temperature at which the reaction occurs, and whether or not any catalysts are present in the reaction.

Gorban and Yablonsky have suggested that the history of chemical dynamics can be divided into three eras. The first is the van 't Hoff wave searching for the general laws of chemical reactions and relating kinetics to thermodynamics. The second may be called the Semenov--Hinshelwood wave with emphasis on reaction mechanisms, especially for chain reactions. The third is associated with Aris and the detailed mathematical description of chemical reaction networks.

Factors Affecting Reaction Rate

Nature of the Reactants

Depending upon what substances are reacting, the reaction rate varies. Acid/base reactions, the formation of salts, and ion exchange are fast reactions. When covalent bond formation takes place between the molecules and when large molecules are formed, the reactions tend to be very slow. Nature and strength of bonds in reactant molecules greatly influence the rate of its transformation into products.

Physical State

The physical state (solid, liquid, or gas) of a reactant is also an important factor of the rate of change. When reactants are in the same phase, as in aqueous solution, thermal motion brings them into contact. However, when they are in different phases, the reaction is limited to the interface between the reactants. Reaction can occur only at their area of contact; in the case of a liquid and a gas, at the surface of the liquid. Vigorous shaking and stirring may be needed to bring the reaction to completion. This means that the more finely divided a solid or liquid reactant the greater its surface area per unit volume and the more contact it with the other reactant, thus the faster the reaction. To make an analogy, for example, when one starts a fire, one uses wood chips and small branches — one does not start with large logs right away. In organic chemistry, on water reactions are the exception to the rule that homogeneous reactions take place faster than heterogeneous reactions.

Surface Area of Solids

In a solid, only those particles that are at the surface can be involved in a reaction. Crushing a solid into smaller parts means that more particles are present at the surface, and the frequency of collisions between these and reactant particles increases, and so reaction occurs more rapidly. For example, Sherbet (powder) is a mixture of very fine powder of malic acid (a weak organic acid) and sodium hydrogen carbonate. On contact with the saliva in the mouth, these chemicals quickly dissolve and react, releasing carbon dioxide and providing for the fizzy sensation. Also, fireworks manufacturers modify the surface area of solid reactants to control the rate at which the fuels in fireworks are oxidised, using this to create different effects. For example, finely divided aluminium confined in a shell explodes violently. If larger pieces of aluminium are used, the reaction is slower and sparks are seen as pieces of burning metal are ejected.

Concentration

The reactions are due to collisions of reactant species. The frequency with which the molecules or ions collide depends upon their concentrations. The more crowded the molecules are, the more likely they are to collide and react with one another. Thus, an increase in the concentrations of the reactants will usually result in the corresponding increase in the reaction rate, while a decrease in the concentrations will usually have a reverse effect. For example, combustion will occur more rapidly in pure oxygen than in air (21% oxygen).

The rate equation shows the detailed dependence of the reaction rate on the concentrations of reactants and other species present. Different mathematical forms are possible depending on the reaction mechanism. The actual rate equation for a given reaction is determined experimentally and provides information about the reaction mechanism.

Temperature

Temperature usually has a major effect on the rate of a chemical reaction. Molecules at a higher temperature have more thermal energy. Although collision frequency is greater at higher temperatures, this alone contributes only a very small proportion to the increase in rate of reaction. Much more important is the fact that the proportion of reactant molecules with sufficient energy to react (energy greater than activation energy: $E > E_a$) is significantly higher and is explained in detail by the Maxwell–Boltzmann distribution of molecular energies.

The 'rule of thumb' that the rate of chemical reactions doubles for every 10 °C temperature rise is a common misconception. This may have been generalized from the special case of biological systems, where the α (temperature coefficient) is often between 1.5 and 2.5.

A reaction's kinetics can also be studied with a temperature jump approach. This involves using a sharp rise in temperature and observing the relaxation time of the return to equilibrium. A particularly useful form of temperature jump apparatus is a shock tube, which can rapidly jump a gas's temperature by more than 1000 degrees.

Catalysts

A catalyst is a substance that alters the rate of a chemical reaction but remains chemically unchanged afterwards. The catalyst increases the rate of the reaction by providing a different reaction mechanism to occur with a lower activation energy. In autocatalysis a reaction product is itself a catalyst for that reaction leading to positive feedback. Proteins that act as catalysts in biochemical reactions are called enzymes. Michaelis–Menten kinetics describe the rate of enzyme mediated reactions. A catalyst does not affect the position of the equilibrium, as the catalyst speeds up the backward and forward reactions equally.

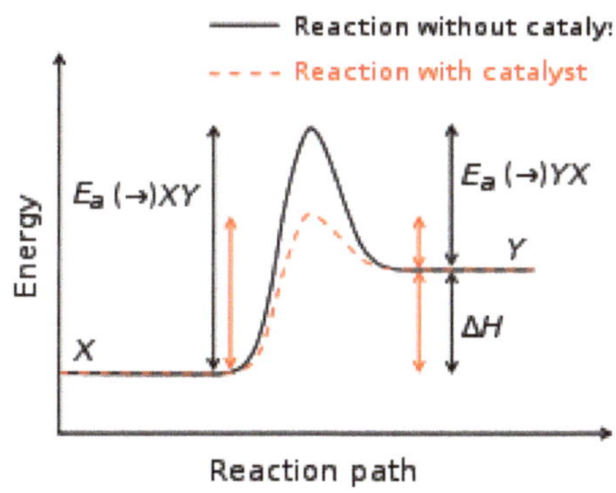

Generic potential energy diagram showing the effect of a catalyst in a hypothetical endothermic chemical reaction. The presence of the catalyst opens a different reaction pathway (shown in red) with a lower activation energy. The final result and the overall thermodynamics are the same.

In certain organic molecules, specific substituents can have an influence on reaction rate in neighbouring group participation.

Pressure

Increasing the pressure in a gaseous reaction will increase the number of collisions between reactants, increasing the rate of reaction. This is because the activity of a gas is directly proportional to the partial pressure of the gas. This is similar to the effect of increasing the concentration of a solution.

In addition to this straightforward mass-action effect, the rate coefficients themselves can change due to pressure. The rate coefficients and products of many high-temperature gas-phase reactions change if an inert gas is added to the mixture; variations on this effect are called fall-off and chemical activation. These phenomena are due to exothermic or endothermic reactions occurring faster than heat transfer, causing the reacting molecules to have non-thermal energy distributions (non-Boltzmann distribution). Increasing the pressure increases the heat transfer rate between the reacting molecules and the rest of the system, reducing this effect.

Condensed-phase rate coefficients can also be affected by (very high) pressure; this is a completely different effect than fall-off or chemical-activation. It is often studied using diamond anvils.

A reaction's kinetics can also be studied with a pressure jump approach. This involves making fast changes in pressure and observing the relaxation time of the return to equilibrium.

Experimental Methods

The experimental determination of reaction rates involves measuring how the concentrations of reactants or products change over time. For example, the concentration of a reactant can be measured by spectrophotometry at a wavelength where no other reactant or product in the system absorbs light.

For reactions which take at least several minutes, it is possible to start the observations after the reactants have been mixed at the temperature of interest.

Fast Reactions

For faster reactions, the time required to mix the reactants and bring them to a specified temperature may be comparable or longer than the half-life of the reaction. Special methods to start fast reactions without slow mixing step include

- Stopped flow methods, which can reduce the mixing time to the order of a millisecond

- Chemical relaxation methods such as temperature jump and pressure jump, in which a pre-mixed system initially at equilibrium is perturbed by rapid heating or depressurization so that it is no longer at equilibrium, and the relaxation back to equilibrium is observed. For example, this method has been used to study the neutralization $H_3O^+ + OH^-$ with a half-life of 1 μs or less under ordinary conditions.
- Flash photolysis, in which a laser pulse produces highly excited species such as free radicals, whose reactions are then studied.

Equilibrium

While chemical kinetics is concerned with the rate of a chemical reaction, thermodynamics determines the extent to which reactions occur. In a reversible reaction, chemical equilibrium is reached when the rates of the forward and reverse reactions are equal (the principle of detailed balance) and the concentrations of the reactants and products no longer change. This is demonstrated by, for example, the Haber–Bosch process for combining nitrogen and hydrogen to produce ammonia. Chemical clock reactions such as the Belousov–Zhabotinsky reaction demonstrate that component concentrations can oscillate for a long time before finally attaining the equilibrium.

Free Energy

In general terms, the free energy change (ΔG) of a reaction determines whether a chemical change will take place, but kinetics describes how fast the reaction is. A reaction can be very exothermic and have a very positive entropy change but will not happen in practice if the reaction is too slow. If a reactant can produce two different products, the thermodynamically most stable one will in general form, except in special circumstances when the reaction is said to be under kinetic reaction control. The Curtin–Hammett principle applies when determining the product ratio for two reactants interconverting rapidly, each going to a different product. It is possible to make predictions about reaction rate constants for a reaction from free-energy relationships.

The kinetic isotope effect is the difference in the rate of a chemical reaction when an atom in one of the reactants is replaced by one of its isotopes.

Chemical kinetics provides information on residence time and heat transfer in a chemical reactor in chemical engineering and the molar mass distribution in polymer chemistry.

Applications and Models

The mathematical models that describe chemical reaction kinetics provide chemists and chemical engineers with tools to better understand and describe chemical processes such as food decomposition, microorganism growth, stratospheric ozone decomposition, and the complex chemistry of biological systems. These models can also be used in the design or modification of chemical reactors to optimize product yield, more efficiently separate products, and eliminate environmentally harmful by-products. When performing catalytic cracking of heavy hydrocarbons into gasoline and light gas, for example, kinetic models can be used to find the temperature and pressure at which the highest yield of heavy hydrocarbons into gasoline will occur.

Chemical Kinetics is frequently validated and explored through modeling in specialized packages as a function of ordinary differential equation-solving (ODE-solving) and curve-fitting.

References

- Howard M. Berlin, Frank C. Getz, Principles of Electronic Instrumentation and Measurement, p. 37, Merrill Pub. Co., 1988 ISBN 0-675-20449-6.
- N. N. Bhargava; D. C. Kulshreshtha (1983). Basic Electronics & Linear Circuits. Tata McGraw-Hill Education. p. 90. ISBN 978-0-07-451965-3.
- Andrew J. Robinson; Lynn Snyder-Mackler (2007). Clinical Electrophysiology: Electrotherapy and Electrophysiologic Testing (3rd ed.). Lippincott Williams & Wilkins. p. 10. ISBN 978-0-7817-4484-3.
- Rudolf Holze, Experimental Electrochemistry: A Laboratory Textbook, page 44, John Wiley & Sons, 2009 ISBN 3527310983.
- Avadhanulu, M.N.; P.G. Kshirsagar (1992). A Textbook Of Engineering Physics For B.E., B.Sc. S. Chand. pp. 345–348. ISBN 8121908175.
- Poole, Charles P. Jr. (2004). Encyclopedic Dictionary of Condensed Matter Physics, Vol. 1. Academic Press. p. 468. ISBN 0080545238.
- Flesch, Peter G. (2007). Light and Light Sources: High-Intensity Discharge Lamps. Springer. pp. 102–103. ISBN 3540326855.
- Ferris, Clifford "Electron tube fundamentals" in Whitaker, Jerry C. (2013). The Electronics Handbook, 2nd Ed. CRC Press. pp. 354–356. ISBN 1420036661.
- Jones, Martin Hartley (1995). A Practical Introduction to Electronic Circuits. UK: Cambridge Univ. Press. p. 49. ISBN 0521478790.
- Sisodia, M. L. (2006). Microwave Active Devices Vacuum and Solid State. New Age International. p. 2.5. ISBN 8122414478.
- McNaught, A. D.; Wilkinson, A. (1997). IUPAC Compendium of Chemical Terminology (2nd ed.). Oxford: Blackwell Scientific Publications. doi:10.1351/goldbook.A00370. ISBN 0-9678550-9-8.
- Espenson, J.H. Chemical Kinetics and Reaction Mechanisms (2nd ed., McGraw-Hill 2002), p.254-256 ISBN 0-07-288362-6
- Steinfeld J.I., Francisco J.S. and Hase W.L. Chemical Kinetics and Dynamics (2nd ed., Prentice-Hall 1999) p.140-3 ISBN 0-13-737123-3
- Steinfeld J.I., Francisco J.S. and Hase W.L. Chemical Kinetics and Dynamics (2nd ed., Prentice-Hall 1999) p.94-97 ISBN 0-13-737123-3

Devices used in Electrochemistry

A number of devices are used in electrochemistry. Some of these devices are electrochemical cells, galvanic cells, thermogalvanic cells, electrolytic cells, fuel cells and photocathodes. An electrochemical cell is a device that is capable of producing electrical energy from chemical reactions whereas galvanic cells are electrochemical cells that develop electrical energy from automatic redox reactions within a cell. This section will provide an integrated understanding of electrochemistry.

Electrochemical Cell

An electrochemical cell is a device capable of either generating electrical energy from chemical reactions or facilitating chemical reactions through the introduction of electrical energy. A common example of an electrochemical cell is a standard 1.5-volt cell meant for consumer use. This type of device is known as a single Galvanic cell. A *battery* consists of two or more cells, connected in either parallel or series pattern.

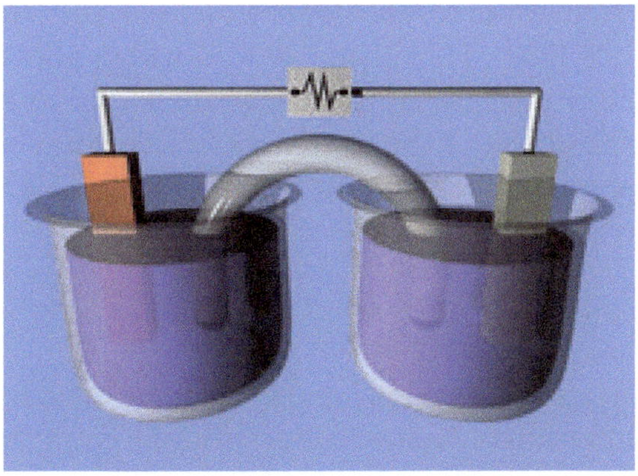

A demonstration electrochemical cell setup resembling the Daniell cell. The two half-cells are linked by a salt bridge carrying ions between them. Electrons flow in the external circuit.

Half-cells

An electrochemical cell consists of two half-cells. Each *half-cell* consists of an electrode and an electrolyte. The two half-cells may use the same electrolyte, or they may use different electrolytes. The chemical reactions in the cell may involve the electrolyte, the electrodes, or an external substance (as in fuel cells that may use hydrogen gas as a reactant). In a full electrochemical cell, species from one half-cell lose electrons (oxidation) to their electrode while species from the other half-cell gain electrons (reduction) from their electrode.

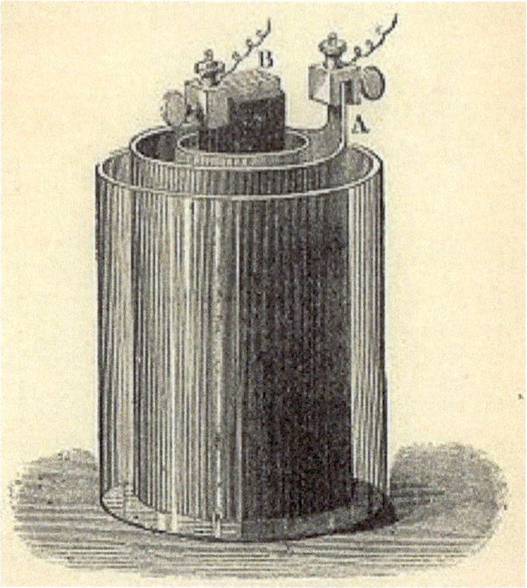

The Bunsen cell, invented by Robert Bunsen.

A *Salt Bridge* (e.g., filter paper soaked in $KNO_{3,}$ NaCl, or some other electrolyte) is often employed to provide ionic contact between two half-cells with different electrolytes, yet prevent the solutions from mixing and causing unwanted side reactions. An alternative to a salt bridge is to allow direct contact (and mixing) between the two half-cells, for example in simple electrolysis of water.

As electrons flow from one half-cell to the other through an external circuit, a difference in charge is established. If no ionic contact were provided, this charge difference would quickly prevent the further flow of electrons. A salt bridge allows the flow of negative or positive ions to maintain a steady-state charge distribution between the oxidation and reduction vessels, while keeping the contents otherwise separate. Other devices for achieving separation of solutions are porous pots and gelled solutions. A porous pot is used in the Bunsen cell (right).

Equilibrium Reaction

Each half-cell has a characteristic voltage. Various choices of substances for each half-cell give different potential differences. Each reaction is undergoing an equilibrium reaction between different oxidation states of the ions: When equilibrium is reached, the cell cannot provide further voltage. In the half-cell that is undergoing oxidation, the closer the equilibrium lies to the ion/atom with the more positive oxidation state the more potential this reaction will provide. Likewise, in the reduction reaction, the closer the equilibrium lies to the ion/atom with the more *negative* oxidation state the higher the potential.

Cell Potential

The cell potential can be predicted through the use of electrode potentials (the voltages of each half-cell). These half-cell potentials are defined relative to the assignment of

0 volts to the standard hydrogen electrode (SHE). The difference in voltage between electrode potentials gives a prediction for the potential measured. When calculating the difference in voltage, one must first rewrite the half-cell reaction equations to obtain a balanced oxidation-reduction equation.

1. Reverse the reduction reaction with the smallest potential (to create an oxidation reaction/ overall positive cell potential)
2. Half-reactions must be multiplied by integers to achieve electron balance.

Note that the cell potential does not change when the reaction is multiplied by a constant.

Cell potentials have a possible range of roughly zero to 6 volts. Cells using water-based electrolytes are usually limited to cell potentials less than about 2.5 volts, because the very powerful oxidizing and reducing agents that would be required to produce a higher cell potential tend to react with the water. Higher cell potentials are possible with cells using other solvents instead of water. For instance, lithium cells with a voltage of 3 volts are commonly available.

The cell potential depends on the concentration of the reactants, as well as their type. As the cell is discharged, the concentration of the reactants decreases, and the cell potential also decreases.

Galvanic Cell

A galvanic cell, or voltaic cell, named after Luigi Galvani, or Alessandro Volta respectively, is an electrochemical cell that derives electrical energy from spontaneous redox reactions taking place within the cell. It generally consists of two different metals connected by a salt bridge, or individual half-cells separated by a porous membrane.

Volta was the inventor of the voltaic pile, the first electrical battery. In common usage, the word "battery" has come to include a single galvanic cell, but a battery properly consists of multiple cells.

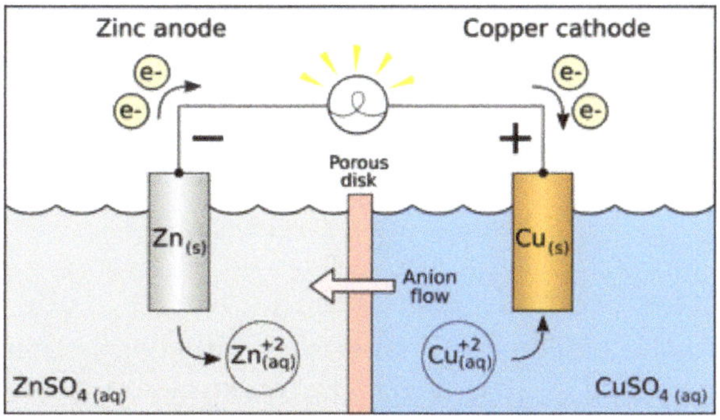

History

In 1780, Luigi Galvani discovered that when two different metals (e.g., copper and zinc) are connected and then both touched at the same time to two different parts of a nerve of a frog leg, then the leg contracts. He called this "animal electricity". The voltaic pile, invented by Alessandro Volta in the 1800s, consists of a pile of cells similar to the galvanic cell. However, Volta built it entirely out of non-biological material in order to challenge Galvani's (and the later experimenter Leopoldo Nobili) animal electricity theory in favour of his own metal-metal contact electricity theory. Carlo Matteucci in his turn constructed a battery entirely out of biological material in answer to Volta. These discoveries paved the way for electrical batteries; Volta's cell was named an IEEE Milestone in 1999.

It was suggested by Wilhelm König in 1940 that the object known as the Baghdad battery might represent galvanic cell technology from ancient Parthia. Replicas filled with citric acid or grape juice have been shown to produce a voltage. However, it is far from certain that this was its purpose—other scholars have pointed out that it is very similar to vessels known to have been used for storing parchment scrolls.

Description

In its simplest form, a half-cell consists of a solid metal (called an electrode) that is submerged in a solution; the solution contains cations of the electrode metal and anions to balance the charge of the cations. In essence, a half-cell contains a metal in two oxidation states; inside an isolated half-cell, there is an oxidation-reduction (redox) reaction that is in chemical equilibrium, a condition written symbolically as follows (here, "M" represents a metal cation, an atom that has a charge imbalance due to the loss of "n" electrons):

$$M^{n+} \text{ (oxidized species)} + ne^- \rightleftharpoons M \text{ (reduced species)}$$

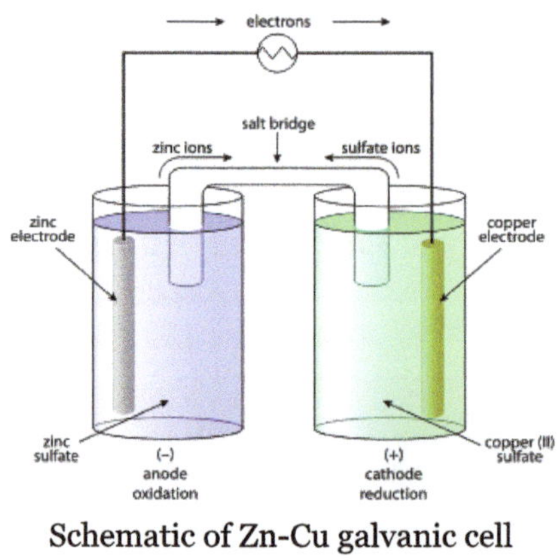

Schematic of Zn-Cu galvanic cell

A galvanic cell consists of two half-cells, such that the electrode of one half-cell is composed of metal A, and the electrode of the other half-cell is composed of metal B; the redox reactions for the two separate half-cells are thus:

$$A^{n+} + ne^- \rightleftharpoons A$$

$$B^{m+} + me^- \rightleftharpoons B$$

In general, then, these two metals can react with each other:

$$m\,A + n\,B^{m+} \rightleftharpoons n\,B + m\,A^{n+}$$

In other words, the metal atoms of one half-cell are able to induce reduction of the metal cations of the other half-cell; conversely stated, the metal cations of one half-cell are able to oxidize the metal atoms of the other half-cell. When metal B has a greater electronegativity than metal A, then metal B tends to steal electrons from metal A (that is, metal B tends to oxidize metal A), thus favoring one direction of the reaction:

$$m\,A + n\,B^{m+} \rightarrow n\,B + m\,A^{n+}$$

This reaction between the metals can be controlled in a way that allows for doing useful work:

- The electrodes are connected with a metal wire in order to conduct the electrons that participate in the reaction.

  In one half-cell, dissolved metal-B cations combine with the free electrons that are available at the interface between the solution and the metal-B electrode; these cations are thereby neutralized, causing them to precipitate from solution as deposits on the metal-B electrode, a process known as plating.

  This reduction reaction causes the free electrons throughout the metal-B electrode, the wire, and the metal-A electrode to be pulled into the metal-B electrode. Consequently, electrons are wrestled away from some of the atoms of the metal-A electrode, as though the metal-B cations were reacting directly with them; those metal-A atoms become cations that dissolve into the surrounding solution.

  As this reaction continues, the half-cell with the metal-A electrode develops a positively charged solution (because the metal-A cations dissolve into it), while the other half-cell develops a negatively charged solution (because the metal-B cations precipitate out of it, leaving behind the anions); unabated, this imbalance in charge would stop the reaction.

- The solutions are connected by a salt bridge or a porous plate in order to conduct the ions (both the metal-A cations from one solution, and the anions from the other solution), which balances the charges of the solutions and thereby allows the reaction between metal A and metal B to continue without opposition.

By definition:

- The anode is the electrode where oxidation (loss of electrons) takes place; in a galvanic cell, it is the negative electrode, as when oxidation occurs, electrons are left behind on the electrode. These electrons then migrate to the cathode (positive electrode). However, in electrolysis, an electric current stimulates electron flow in the opposite direction. Thus, the anode is positive, and the statement anode attracts anions is true (negatively charged ions flow to the anode, while electrons are expelled through the wire). The metal-A electrode is the anode.

- The cathode is the electrode where reduction (gain of electrons) takes place; in a galvanic cell, it is the positive electrode, as less oxidation occurs, fewer ions go into solution, and less electrons are left on the electrode. Instead, there is a greater tendency for aqueous ions to be reduced by the incoming electrons from the anode. However, in electrolysis, the cathode is the negative terminal, and attracting positive ions from the solution. In this situation, the statement the cathode attracts cations is true (positively charged, oxidized metal ions flow toward cathode as electrons travel through the wire). The metal-B electrode is the cathode.

Copper readily oxidizes zinc; for the Daniell cell depicted in the figure, the anode is zinc and the cathode is copper, and the anions in the solutions are sulfates of the respective metals. When an electrically conducting device connects the electrodes, the electrochemical reaction is:

$$Zn + Cu^{2+} \rightarrow Zn^{2+} + Cu$$

The zinc electrode is dissolved and copper is deposited on the copper electrode.

Galvanic cells are typically used as a source of electrical power. By their nature, they produce direct current. The Weston cell has an anode composed of cadmium mercury amalgam, and a cathode composed of pure mercury. The electrolyte is a (saturated) solution of cadmium sulfate. The depolarizer is a paste of mercurous sulfate. When the electrolyte solution is saturated, the voltage of the cell is very reproducible; hence, in 1911, it was adopted as an international standard for voltage.

A battery is a set of galvanic cells that are connected in parallel. For instance, a lead–acid battery has galvanic cells with the anodes composed of lead and cathodes composed of lead dioxide.

Cell Voltage

The standard electrical potential of a cell can be determined by the use of a standard potential table for the two half cells involved. The first step is to identify the two metals reacting in the cell. Then one looks up the standard electrode potential, E^o, in volts, for each of the two half reactions. The standard potential for the cell is equal to the more positive E^o value minus the more negative E^o value.

For example, in the figure above the solutions are $CuSO_4$ and $ZnSO_4$. Each solution has a corresponding metal strip in it, and a salt bridge or porous disk connecting the two solutions and allowing SO_4^{2-} ions to flow freely between the copper and zinc solutions. In order to calculate the standard potential one looks up copper and zinc's half reactions and finds:

$$Cu^{2+} + 2e^- \rightleftharpoons Cu: E^o = +0.34 \text{ V}$$

$$Zn^{2+} + 2e^- \rightleftharpoons Zn: E^o = -0.76 \text{ V}$$

Thus the overall reaction is:

$$Cu^{2+} + Zn \rightleftharpoons Cu + Zn^{2+}$$

The standard potential for the reaction is then +0.34 V – (−0.76 V) = 1.10 V. The polarity of the cell is determined as follows. Zinc metal is more strongly reducing than copper metal; equivalently, the standard (reduction) potential for zinc is more negative than that of copper. Thus, zinc metal will lose electrons to copper ions and develop a positive electrical charge. The equilibrium constant, *K*, for the cell is given by

$$\ln K = \frac{nFE^0}{RT}$$

where *F* is the Faraday constant, *R* is the gas constant and *T* is the temperature in kelvins. For the Daniell cell *K* is approximately equal to 1.5×10^{37}. Thus, at equilibrium, a few electrons are transferred, enough to cause the electrodes to be charged.

Actual half-cell potentials must be calculated by using the Nernst equation as the solutes are unlikely to be in their standard states,

$$E_{\text{half-cell}} = E^0 - \frac{RT}{nF} \ln_e Q$$

where *Q* is the reaction quotient. This simplifies to

$$E_{\text{half-cell}} = E^0 - 2.303 \frac{RT}{nF} \log_{10} \{M^{n+}\}$$

where $\{M^{n+}\}$ is the activity of the metal ion in solution. The metal electrode is in its standard state so by definition has unit activity. In practice concentration is used in place of activity. The potential of the whole cell is obtained by combining the potentials for the two half-cells, so it depends on the concentrations of both dissolved metal ions.

The value of $2.303R/F$ is 0.19845×10^{-3} V/K, so at 25 °C (298.15 K) the half-cell potential will change by $0.05918V >$ if the concentration of a metal ion is increased or decreased by a factor of 10.

$$E_{\text{half-cell}} = E^0 - \frac{0.05918V}{n}\log_{10}[M^{n+}]$$

These calculations are based on the assumption that all chemical reactions are in equilibrium. When a current flows in the circuit, equilibrium conditions are not achieved and the cell potential will usually be reduced by various mechanisms, such as the development of overpotentials. Also, since chemical reactions occur when the cell is producing power, the electrolyte concentrations change and the cell voltage is reduced. A consequence of the temperature dependency of standard potentials is that the voltage produced by a galvanic cell is also temperature dependent.

Galvanic Corrosion

Galvanic corrosion is a process that degrades metals electrochemically. This corrosion occurs when two dissimilar metals are located in contact with each other in the presence of an electrolyte, such as salt water, forming a galvanic cell. A cell can also be formed if the same metal is exposed to two different concentrations of electrolyte. The resulting electrochemical potential then develops an electric current that electrolytically dissolves the less noble material.

Thermogalvanic Cell

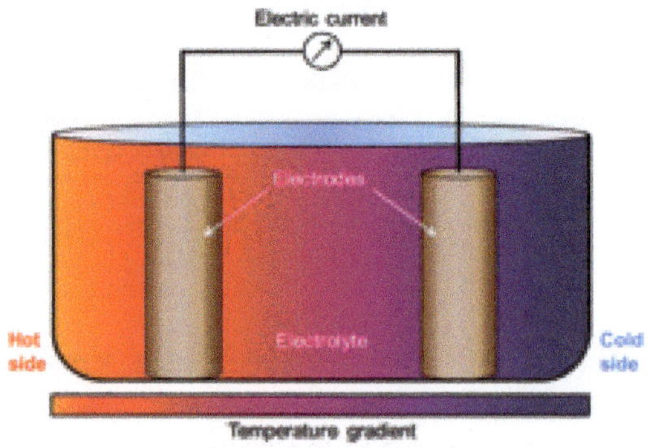

Thermogalvanic cell displaying the elements making up the cell

A thermogalvanic cell is a kind of galvanic cell in which heat is employed to provide electrical power directly. These cells are electrochemical cells in which the two electrodes are deliberately maintained at different temperatures. This temperature difference generates a potential difference between the electrodes. The electrodes can be of identical composition and the electrolyte solution homogeneous. This is usually the case in these cells. This is in contrast to galvanic cells in which electrodes and/or solutions of different composition provide the electromotive potential. As long as there is a difference in temperature between the electrodes a current will flow through the circuit.

A thermogalvanic cell can be seen as analogous to a concentration cell but instead of running on differences in the concentration/pressure of the reactants they make use of differences in the "concentrations" of thermal energy. The principal application of thermogalvanic cells is the production of electricity from low-temperature heat sources (waste heat and solar heat). Their energetic efficiency is low, in the range of 0.1% to 1% for conversion of heat into electricity.

History

The use of heat to empower galvanic cells was first studied around 1880. However it was not until the decade of 1950 that more serious research was undertaken in this field.

Working Mechanism

Thermogalvanic cells are a kind of heat engine. Ultimately the driving force behind them is the transport of entropy from the high temperature source to the low temperature sink. Therefore these cells work thanks to a thermal gradient established between different parts of the cell. Because the rate and enthalpy of chemical reactions depend directly on the temperature, different temperatures at the electrodes imply different chemical equilibrium constants. This tranlates into unequal chemical equilibrium conditions on the hot side and on the cold side. The thermocell tries to approach an homogeneous equilibrium and, in doing so, produces a flow of chemical species and electrons. The electrons flow through the path of least resistance (the outer circuit) making it possible to extract power from the cell.

Types

Different thermogalvanic cells have been constructed attending to their uses and properties. Usually they are classified according to the electrolyte employed in each specific type of cell.

Aqueous Electrolytes

In these cells the electrolyte between the electrodes is a water solution of some salt or hydrophylic compound. An essential property of these compounds is that they must be able to undergo redox reactions in order to shuttle electrons from one electrode to the other during the cell operation.

Non-aqueous Electrolytes

The electrolyte is a solution of some other solvent different from water. Solvents like methanol, acetone, dimethyl sulphoxide and dimethyl formamide have been successfully employed in thermogalvanic cells running on copper sulfate.

Molten Salts

In this type of thermocell the electrolyte is some kind of salt with a relatively low melting point. Their use solve two problems. On one hand the temperature range of the cell is much larger. This is an advantage as these cells produce more power the larger the difference between the hot and cold sides. On the other hand the liquid salt directly provides the anions and cations necessary for sustainment of a current through the cell. Therefore no additional current-carrying compounds are necessary as the melted salt is the electrolyte itself. Typical hot source temperatures are between 600-900 K, but can get as high as 1730 K. Cold sink temperatures are in the 400-500 K range.

Solid Electrolytes

Thermocells in which the electrolyte connecting the electrodes is an ionic material have been considered and constructed too. The temperature range is also elevated as compared to liquid electrolytes. Studied systems fall in the 400-900 K. Some solid ionic materials that have been employed to construct thermogalvanic cells are AgI, $PbCl_2$ and $PbBr_2$.

Uses

Given the advantages provided by the working mechanism of thermogalvanic cells, their main application is electricity production under conditions where there is a excess of heat available. In particular thermogalvanic cells are being used to produce electricity in the following areas.

Electrolytic Cell

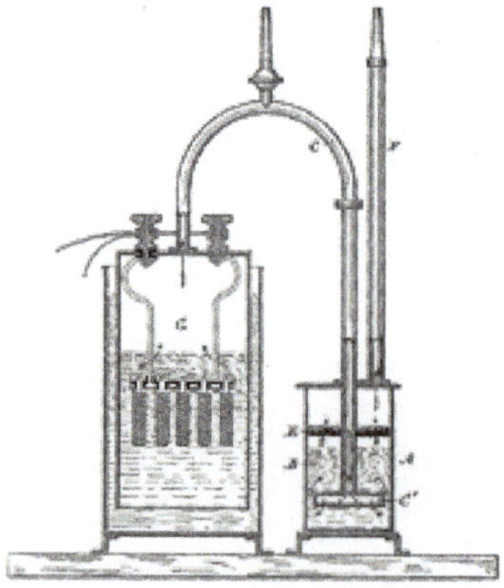

Nineteenth century electrolytic cell for producing oxyhydrogen.

An electrolytic cell is an electrochemical cell that undergoes a redox reaction when electrical energy is applied. It is most often used to decompose chemical compounds, in a process called electrolysis—the Greek word lysis means *to break up*. When electrical energy is added to the system, the chemical energy is increased. Similarly to a galvanic cell, electrolytic cells usually consist of two half cells.

Electrolytic cells for hydrogen production, Fixed Nitrogen Research Laboratory, ca.1930s

Important examples of electrolysis are the decomposition of water into hydrogen and oxygen, and bauxite into aluminium and other chemicals. Electroplating (e.g. of copper, silver, nickel or chromium) is performed using an electrolytic cell.

An electrolytic cell has three component parts: an electrolyte and two electrodes (a cathode and an anode). The electrolyte is usually a solution of water or other solvents in which ions are dissolved. Molten salts such as sodium chloride are also electrolytes. When driven by an external voltage applied to the electrodes, the ions in the electrolyte are attracted to an electrode with the opposite charge, charge-transferring (also called faradaic or redox) reactions can take place. Only with an external electrical potential (i.e. voltage) of correct polarity and sufficient magnitude can an electrolytic cell decompose a normally stable, or inert chemical compound in the solution. The electrical energy provided can produce a chemical reaction which would not occur spontaneously otherwise.

Galvanic Cells Compared to Electrolytic Cells

In contrast, a shaft battery or Galvanic cell, converts chemical energy into electrical energy, by using spontaneous chemical reactions that take place at the electrodes. Each galvanic cell has its own characteristic voltage (defined as the energy release per electron transfer from one electrode to the other). A simple galvanic cell will consist of only an electrolyte and two different electrodes. (Galvanic cells can also be made by connecting two half-cells, each with its own electrode and electrolyte, by an ion-transporting "bridge," usually a salt bridge; these cells are more complex.) The electrodes typically are two metals, which naturally have different reaction potentials relative to the electrolyte. This causes electrons of one of the electrodes to preferentially enter the solution at one electrode, and other electrons to leave the solution at the other electrode. This generates an electric current across the electrolyte (where the charge carriers are the mobile ionic species), which will drive electric current through a wire that makes an exterior connection to each of the electrodes. Both galvanic cells and electrolytic cells can use electrodes of different metals but can also use the same metal for the electrodes.

A rechargeable battery, such as a AA NiMH cell or a single cell of a lead-acid battery, acts as a galvanic cell when discharging (converting chemical energy to electrical energy), and an electrolytic cell when being charged (converting electrical energy to chemical energy).

Anode and Cathode Definitions Depend on Charge and Discharge

Michael Faraday defined the cathode of a cell as the electrode to which cations (positively charged ions, like silver ions Ag+) flow within the cell, to be reduced by reacting with electrons (negatively charged) from that electrode.

Likewise he defined the anode as the electrode to which anions (negatively charged ions, like chloride ions Cl−) flow within the cell, to be oxidized by depositing electrons on the electrode.

To an external wire connected to the electrodes of a Galvanic cell (or battery), forming an electric circuit, the cathode is positive and the anode is negative. Thus positive electric current flows from the cathode to the anode through the external circuit in the case of a Galvanic cell.

Consider two voltaic cells of unequal voltage. Mark the positive and negative electrodes of each one as P and N, respectively. Place them in a circuit with P near P and N near N, so the cells will tend to drive current in opposite directions. The cell with the larger voltage is discharged, making it a galvanic cell, so P is the cathode and N is the anode as described above. But, the cell with the smaller voltage charges, making it an electrolytic cell. In the electrolytic cell, negative ions are driven towards P and positive ions towards N. Thus, the P electrode of the electrolytic cell meets the definition of anode while the electrolytic cell is being charged. Similarly, the N electrode of the electrolytic cell is the cathode while the electrolytic cell is being charged.

Uses

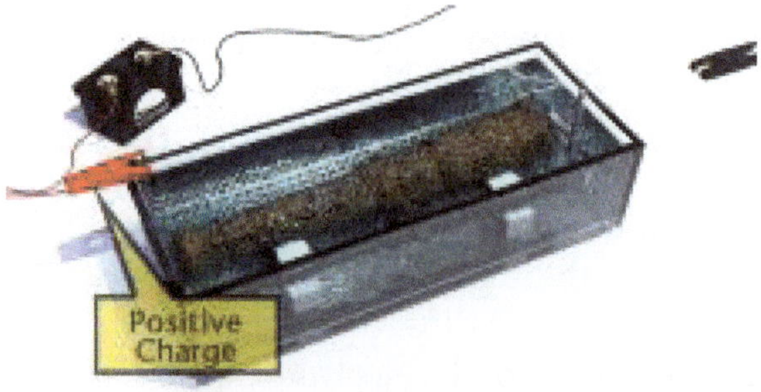

A video describing the process of electrolytic reduction used in a display at The Children's Museum of Indianapolis within the Treasures of the Earth exhibit. The video describes the electrolytic process as it is occurring on Captain Kidd's Cannon, which is on display and currently undergoing the electrolytic process.

As already noted, water, particularly when ions are added (salt water or acidic water), can be *electrolyzed* (subject to electrolysis). When driven by an external source of voltage, H+ions flow to the cathode to combine with electrons to produce hydrogen gas in a reduction reaction. Likewise, OH–ions flow to the anode to release electrons and an H+ion to produce oxygen gas in an oxidation reaction.

In molten sodium chloride, when a current is passed through the salt the anode oxidizes chloride ions (Cl–) to chlorine gas, releasing electrons to the anode. Likewise the cathode reduces sodium ions (Na+), which accept electrons from the cathode and deposits on the cathode as sodium metal.

NaCl dissolved in water can also be electrolyzed. The anode oxidizes chloride ions (Cl–), and Cl_2 gas is produced. However, at the cathode, instead of sodium ions being reduced to sodium metal, water molecules are reduced to hydroxide ions (OH–) and hydrogen gas (H_2). The overall result of the electrolysis is the production of chlorine gas and aqueous sodium hydroxide (NaOH) solution.

Commercially, electrolytic cells are used in electrorefining and electrowinning of several non-ferrous metals. Almost all high-purity aluminium, copper, zinc and lead is produced industrially in electrolytic cells.

Battery (Electricity)

An electric battery is a device consisting of one or more electrochemical cells with external connections provided to power electrical devices such as flashlights, smartphones, and electric cars. When a battery is supplying electric power, its positive terminal is the cathode and its negative terminal is the anode. The terminal marked negative is the source of electrons that when connected to an external circuit will flow and deliver energy to an external device. When a battery is connected to an external circuit, electrolytes are able to move as ions within, allowing the chemical reactions to be completed at the separate terminals and so deliver energy to the external circuit. It is the movement of those ions within the battery which allows current to flow out of the battery to perform work. Historically the term "battery" specifically referred to a device composed of multiple cells, however the usage has evolved to additionally include devices composed of a single cell.

Primary (single-use or "disposable") batteries are used once and discarded; the electrode materials are irreversibly changed during discharge. Common examples are the alkaline battery used for flashlights and a multitude of portable electronic devices. Secondary (rechargeable) batteries can be discharged and recharged multiple times using mains power from a wall socket; the original composition of the electrodes can be restored by reverse current. Examples include the lead-acid batteries used in vehicles and lithium-ion batteries used for portable electronics such as laptops and smartphones.

Batteries come in many shapes and sizes, from miniature cells used to power hearing aids and wristwatches to small, thin cells used in smartphones, to large lead acid batteries used in cars and trucks, and at the largest extreme, huge battery banks the size of rooms that provide standby or emergency power for telephone exchanges and computer data centers.

According to a 2005 estimate, the worldwide battery industry generates US$48 billion in sales each year, with 6% annual growth.

Batteries have much lower specific energy (energy per unit mass) than common fuels such as gasoline. This is somewhat offset by the higher efficiency of electric motors in producing mechanical work, compared to combustion engines.

History

The usage of "battery" to describe a group of electrical devices dates to Benjamin Franklin, who in 1748 described multiple Leyden jars by analogy to a battery of cannon (Benjamin Franklin borrowed the term "battery" from the military, which refers to weapons functioning together).

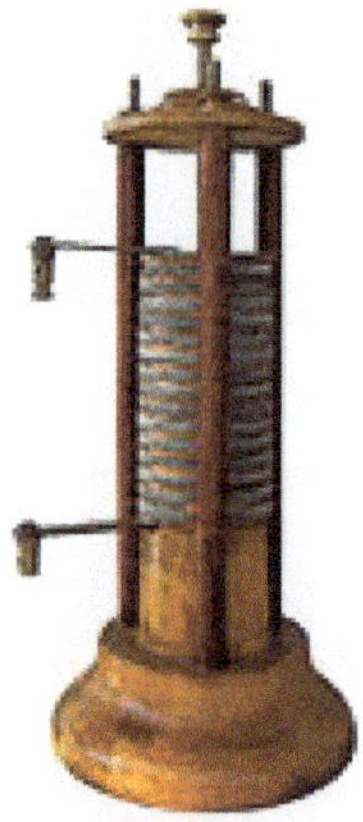

A voltaic pile, the first battery

Alessandro Volta demonstrating his pile to French emperor Napoleon Bonaparte

Alessandro Volta built and described the first electrochemical battery, the voltaic pile, in 1800. This was a stack of copper and zinc plates, separated by brine-soaked paper disks, that could produce a steady current for a considerable length of time. Volta did not understand that the voltage was due to chemical reactions. He thought that his cells were an inexhaustible source of energy, and that the associated corrosion effects at the electrodes were a mere nuisance, rather than an unavoidable consequence of their operation, as Michael Faraday showed in 1834.

Although early batteries were of great value for experimental purposes, in practice their voltages fluctuated and they could not provide a large current for a sustained period. The Daniell cell, invented in 1836 by British chemist John Frederic Daniell, was the first practical source of electricity, becoming an industry standard and seeing widespread adoption as a power source for electrical telegraph networks. It consisted of a copper pot filled with a copper sulfate solution, in which was immersed an unglazed earthenware container filled with sulfuric acid and a zinc electrode.

These wet cells used liquid electrolytes, which were prone to leakage and spillage if not handled correctly. Many used glass jars to hold their components, which made them fragile and potentially dangerous. These characteristics made wet cells unsuitable for portable appliances. Near the end of the nineteenth century, the invention of dry cell batteries, which replaced the liquid electrolyte with a paste, made portable electrical devices practical.

Principle of Operation

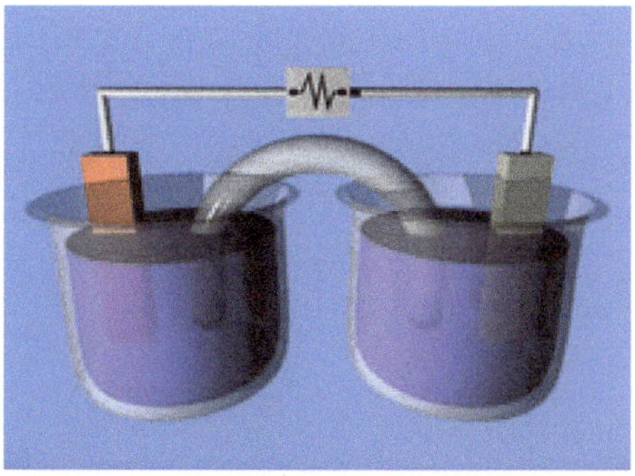

A voltaic cell for demonstration purposes. In this example the two half-cells are linked by a salt bridge separator that permits the transfer of ions.

Batteries convert chemical energy directly to electrical energy. A battery consists of some number of voltaic cells. Each cell consists of two half-cells connected in series by a conductive electrolyte containing anions and cations. One half-cell includes electrolyte and the negative electrode, the electrode to which anions (negatively charged ions) migrate; the other half-cell includes electrolyte and the positive electrode to which cations (positively charged ions) migrate. Redox reactions power the battery. Cations are reduced (electrons are added) at the cathode during charging, while anions are oxidized (electrons are removed) at the anode during charging. During discharge, the process is reversed. The electrodes do not touch each other, but are electrically connected by the electrolyte. Some cells use different electrolytes for each half-cell. A separator allows ions to flow between half-cells, but prevents mixing of the electrolytes.

Each half-cell has an electromotive force (or emf), determined by its ability to drive electric current from the interior to the exterior of the cell. The net emf of the cell is the difference between the emfs of its half-cells. Thus, if the electrodes have emfs and , then the net emf is ; in other words, the net emf is the difference between the reduction potentials of the half-reactions.

The electrical driving force or across the terminals of a cell is known as the *terminal voltage (difference)* and is measured in volts. The terminal voltage of a cell that is neither charging nor discharging is called the open-circuit voltage and equals the emf of the cell. Because of internal resistance, the terminal voltage of a cell that is discharging is smaller in magnitude than the open-circuit voltage and the terminal voltage of a cell that is charging exceeds the open-circuit voltage. An ideal cell has negligible internal resistance, so it would maintain a constant terminal voltage of until exhausted, then dropping to zero. If such a cell maintained 1.5 volts and stored a charge of one coulomb then on complete discharge it would perform 1.5 joules of work. In actual cells, the internal resistance increases under discharge and the open circuit voltage also decreases under discharge. If the voltage and resistance are plotted against time, the resulting graphs typically are a curve; the shape of the curve varies according to the chemistry and internal arrangement employed.

The voltage developed across a cell's terminals depends on the energy release of the chemical reactions of its electrodes and electrolyte. Alkaline and zinc–carbon cells have different chemistries, but approximately the same emf of 1.5 volts; likewise NiCd and NiMH cells have different chemistries, but approximately the same emf of 1.2 volts. The high electrochemical potential changes in the reactions of lithium compounds give lithium cells emfs of 3 volts or more.

Categories and Types of Batteries

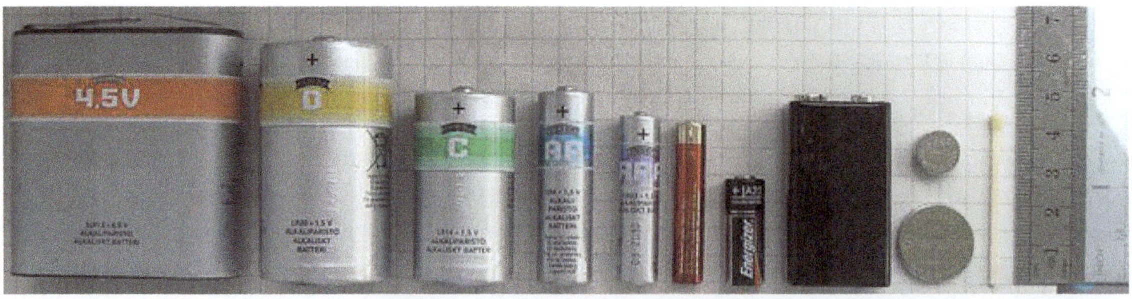

From left to right: a large 4.5-volt (3R12) battery, a *D Cell*, a *C cell*, an *AA cell*, an *AAA cell*, an *AAAA cell*, an *A23 battery*, a 9-volt *PP3 battery*, and a pair of button cells (CR2032 and LR44)

Batteries are classified into primary and secondary forms:

- *Primary* batteries are designed to be used until exhausted of energy then discarded. Their chemical reactions are generally not reversible, so they cannot be recharged. When the supply of reactants in the battery is exhausted, the battery stops producing current and is useless.
- *Secondary* batteries can be recharged; that is, they can have their chemical reactions reversed by applying electric current to the cell. This regenerates the original chemical reactants, so they can be used, recharged, and used again multiple times.

Some types of primary batteries used, for example, for telegraph circuits, were restored to operation by replacing the electrodes. Secondary batteries are not indefinitely rechargeable due to dissipation of the active materials, loss of electrolyte and internal corrosion.

Primary

Primary batteries, or primary cells, can produce current immediately on assembly. These are most commonly used in portable devices that have low current drain, are used only intermittently, or are used well away from an alternative power source, such as in alarm and communication circuits where other electric power is only intermittently available. Disposable primary cells cannot be reliably recharged, since the chemical reactions are not easily reversible and active materials may not return to their original forms. Battery manufacturers recommend against attempting to recharge primary cells. In general, these have higher energy densities than rechargeable batteries, but disposable batteries do not fare well under high-drain applications with loads under 75 ohms (75 Ω). Common types of disposable batteries include zinc–carbon batteries and alkaline batteries.

Secondary

Secondary batteries, also known as *secondary cells*, or *rechargeable batteries*, must be charged before first use; they are usually assembled with active materials in the discharged state. Rechargeable batteries are (re)charged by applying electric current, which reverses the chemical reactions that occur during discharge/use. Devices to supply the appropriate current are called chargers.

The oldest form of rechargeable battery is the lead–acid battery, which are widely used in automotive and boating applications. This technology contains liquid electrolyte in an unsealed container, requiring that the battery be kept upright and the area be well ventilated to ensure safe dispersal of the hydrogen gas it produces during overcharging. The lead–acid battery is relatively heavy for the amount of electrical energy it can supply. Its low manufacturing cost and its high surge current levels make it common where its capacity (over approximately 10 Ah) is more important than weight and handling issues. A common application is the modern car battery, which can, in general, deliver a peak current of 450 amperes.

The sealed valve regulated lead–acid battery (VRLA battery) is popular in the automotive industry as a replacement for the lead–acid wet cell. The VRLA battery uses an immobilized sulfuric acid electrolyte, reducing the chance of leakage and extending shelf life. VRLA batteries immobilize the electrolyte. The two types are:

- *Gel batteries* (or "gel cell") use a semi-solid electrolyte.
- *Absorbed Glass Mat* (AGM) batteries absorb the electrolyte in a special fiberglass matting.

Other portable rechargeable batteries include several sealed "dry cell" types, that are useful in applications such as mobile phones and laptop computers. Cells of this type (in order of increasing power density and cost) include nickel–cadmium (NiCd), nickel–zinc (NiZn), nickel metal hydride (NiMH), and lithium-ion (Li-ion) cells. Li-ion has by far the highest share of the dry cell rechargeable market. NiMH has replaced NiCd in most applications due to its higher capacity, but NiCd remains in use in power tools, two-way radios, and medical equipment.

In the 2000s, developments include batteries with embedded electronics such as USBCELL, which allows charging an AA battery through a USB connector, nanoball batteries that allow for a discharge rate about 100x greater than current batteries, and smart battery packs with state-of-charge monitors and battery protection circuits that prevent damage on over-discharge. Low self-discharge (LSD) allows secondary cells to be charged prior to shipping.

Cell Types

Many types of electrochemical cells have been produced, with varying chemical processes and designs, including galvanic cells, electrolytic cells, fuel cells, flow cells and voltaic piles.

Wet Cell

A *wet cell* battery has a liquid electrolyte. Other names are *flooded cell*, since the liquid covers all internal parts, or *vented cell*, since gases produced during operation can escape to the air. Wet cells were a precursor to dry cells and are commonly used as a learning tool for electrochemistry. They can be built with common laboratory supplies, such as beakers, for demonstrations of how electrochemical cells work. A particular type of wet cell known as a concentration cell is important in understanding corrosion. Wet cells may be primary cells (non-rechargeable) or secondary cells (rechargeable). Originally, all practical primary batteries such as the Daniell cell were built as open-top glass jar wet cells. Other primary wet cells are the Leclanche cell, Grove cell, Bunsen cell, Chromic acid cell, Clark cell, and Weston cell. The Leclanche cell chemistry was adapted to the first dry cells. Wet cells are still used in automobile batteries and in industry for standby power for switchgear, telecommunication or large uninterruptible power supplies, but in many places batteries with gel cells have been used instead. These applications commonly use lead–acid or nickel–cadmium cells.

Dry Cell

A *dry cell* uses a paste electrolyte, with only enough moisture to allow current to flow. Unlike a wet cell, a dry cell can operate in any orientation without spilling, as it contains no free liquid, making it suitable for portable equipment. By comparison, the first wet cells were typically fragile glass containers with lead rods hanging from the open top and needed careful handling to avoid spillage. Lead–acid batteries did not achieve the safety and portability of the dry cell until the development of the gel battery.

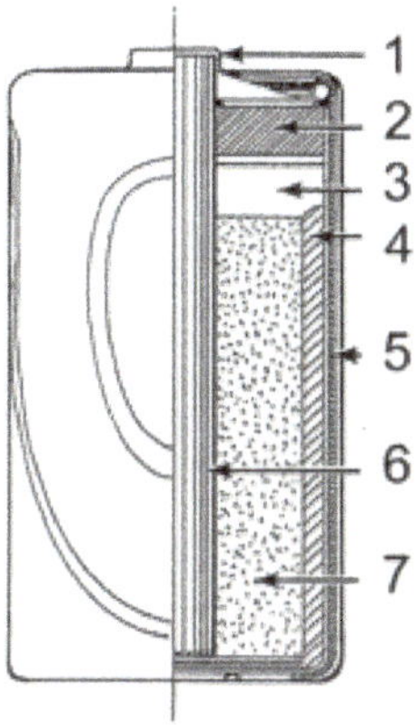

Line art drawing of a dry cell: 1. brass cap, 2. plastic seal, 3. expansion space, 4. porous cardboard, 5. zinc can, 6. carbon rod, 7. chemical mixture

A common dry cell is the zinc–carbon battery, sometimes called the dry Leclanché cell, with a nominal voltage of 1.5 volts, the same as the alkaline battery (since both use the same zinc–manganese dioxide combination). A standard dry cell comprises a zinc anode, usually in the form of a cylindrical pot, with a carbon cathode in the form of a central rod. The electrolyte is ammonium chloride in the form of a paste next to the zinc anode. The remaining space between the electrolyte and carbon cathode is taken up by a second paste consisting of ammonium chloride and manganese dioxide, the latter acting as a depolariser. In some designs, the ammonium chloride is replaced by zinc chloride.

Molten Salt

Molten salt batteries are primary or secondary batteries that use a molten salt as electrolyte. They operate at high temperatures and must be well insulated to retain heat.

Reserve

A reserve battery can be stored unassembled (unactivated and supplying no power) for a long period (perhaps years). When the battery is needed, then it is assembled (e.g., by adding electrolyte); once assembled, the battery is charged and ready to work. For example, a battery for an electronic artillery fuze might be activated by the impact of firing a gun: The acceleration breaks a capsule of electrolyte that activates the battery and powers the fuze's circuits. Reserve batteries are usually designed for a short service life (seconds or minutes) after long storage (years). A water-activated battery for oceanographic instruments or military applications becomes activated on immersion in water.

Cell Performance

A battery's characteristics may vary over load cycle, over charge cycle, and over lifetime due to many factors including internal chemistry, current drain, and temperature. At low temperatures, a battery cannot deliver as much power. As such, in cold climates, some car owners install battery warmers, which are small electric heating pads that keep the car battery warm.

Capacity and Discharge

A battery's *capacity* is the amount of electric charge it can deliver at the rated voltage. The more electrode material contained in the cell the greater its capacity. A small cell has less capacity than a larger cell with the same chemistry, although they develop the same open-circuit voltage. Capacity is measured in units such as amp-hour (A·h). The rated capacity of a battery is usually expressed as the product of 20 hours multiplied by the current that a new battery can consistently supply for 20 hours at 68 °F (20 °C), while remaining above a specified terminal voltage per cell. For example, a battery rated at 100 A·h can deliver 5 A over a 20-hour period at room temperature. The fraction of the stored charge that a battery can deliver depends on multiple factors, including battery chemistry, the rate at which the charge is delivered (current), the required terminal voltage, the storage period, ambient temperature and other factors.

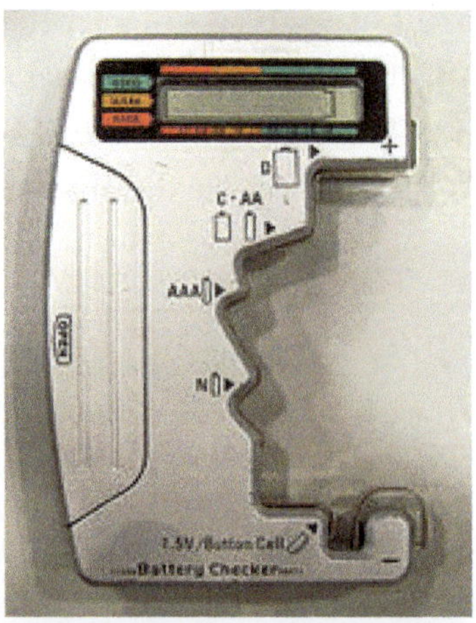

A device to check battery voltage

The higher the discharge rate, the lower the capacity. The relationship between current, discharge time and capacity for a lead acid battery is approximated (over a typical range of current values) by Peukert's law:

$$t = \frac{Q_P}{I^k}$$

where

Q_P is the capacity when discharged at a rate of 1 amp.

I is the current drawn from battery (A).

t is the amount of time (in hours) that a battery can sustain.

k is a constant around 1.3.

Batteries that are stored for a long period or that are discharged at a small fraction of the capacity lose capacity due to the presence of generally irreversible *side reactions* that consume charge carriers without producing current. This phenomenon is known as internal self-discharge. Further, when batteries are recharged, additional side reactions can occur, reducing capacity for subsequent discharges. After enough recharges, in essence all capacity is lost and the battery stops producing power.

Internal energy losses and limitations on the rate that ions pass through the electrolyte cause battery efficiency to vary. Above a minimum threshold, discharging at a low rate delivers more of the battery's capacity than at a higher rate. Installing batteries with varying A·h ratings does not affect device operation (although it may affect the operation interval) rated for a specific voltage unless load limits are exceeded. High-drain loads such as digital cameras can reduce total capacity, as happens with alkaline batteries. For example, a battery rated at 2 A·h for a 10- or 20-hour discharge would not sustain a current of 1 A for a full two hours as its stated capacity implies.

C Rate

The C-rate is a measure of the rate at which a battery is being discharged. It is defined as the discharge current divided by the theoretical current draw under which the battery would deliver its nominal rated capacity in one hour. A 1C discharge rate would deliver the battery's rated capacity in 1 hour. A 2C discharge rate means it will discharge twice as fast (30 minutes). A 1C discharge rate on a 1.6 Ah battery means a discharge current of 1.6 A. A 2C rate would mean a discharge current of 3.2 A. Standards for rechargeable batteries generally rate the capacity over a 4-hour, 8 hour or longer discharge time. Because of internal resistance loss and the chemical processes inside the cells, a battery rarely delivers nameplate rated capacity in only one hour. Types intended for special purposes, such as in a computer uninterruptible power supply, may be rated by manufacturers for discharge periods much less than one hour.

Fast-charging, Large and Light Batteries

As of 2012, lithium iron phosphate (LiFePO4) battery technology was the fastest-charging/discharging, fully discharging in 10–20 seconds.

As of 2013, the world's largest battery was in Hebei Province, China. It stored 36 megawatt-hours of electricity at a cost of $500 million. Another large battery, composed of Ni–Cd cells, was in Fairbanks, Alaska. It covered 2,000 square metres (22,000 sq ft)—bigger than a football pitch—and weighed 1,300 tonnes. It was manufactured by ABB to provide backup power in the event of a blackout. The battery can provide 40 megawatts of power for up to seven minutes. Sodium–sulfur batteries have been used to store wind power. A 4.4 megawatt-hour battery system that can deliver 11 megawatts for 25 minutes stabilizes the output of the Auwahi wind farm in Hawaii.

Lithium–sulfur batteries were used on the longest and highest solar-powered flight. The recharging speed of lithium-ion batteries can be increased by manufacturing changes.

Lifetime

Battery life (and its synonym battery lifetime) has two meanings for rechargeable batteries but only one for non-chargeables. For rechargeables, it can mean either the length of time a device can run on a fully charged battery or the number of charge/discharge cycles possible before the cells fail to operate satisfactorily. For a non-rechargeable these two lives are equal since the cells last for only one cycle by definition. (The term shelf life is used to describe how long a battery will retain its performance between manufacture and use.) Available capacity of all batteries drops with decreasing temperature. In contrast to most of today's batteries, the Zamboni pile, invented in 1812, offers a very long service life without refurbishment or recharge, although it supplies current only in the nanoamp range. The Oxford Electric Bell has been ringing almost continuously since 1840 on its original pair of batteries, thought to be Zamboni piles.

Self-discharge

Disposable batteries typically lose 8 to 20 percent of their original charge per year when stored at room temperature (20–30 °C). This is known as the "self-discharge" rate, and is due to non-current-producing "side" chemical reactions that occur within the cell even when no load is applied. The rate of side reactions is reduced for batteries are stored at lower temperatures, although some can be damaged by freezing.

Old rechargeable batteries self-discharge more rapidly than disposable alkaline batteries, especially nickel-based batteries; a freshly charged nickel cadmium (NiCd) battery loses 10% of its charge in the first 24 hours, and thereafter discharges at a rate of about 10% a month. However, newer low self-discharge nickel metal hydride (NiMH) batteries and modern lithium designs display a lower self-discharge rate (but still higher than for primary batteries).

Corrosion

Internal parts may corrode and fail, or the active materials may be slowly converted to inactive forms.

Physical Component Changes

The active material on the battery plates changes chemical composition on each charge and discharge cycle; active material may be lost due to physical changes of volume, further limiting the number of times the battery can be recharged. Most nickel-based batteries are partially discharged when purchased, and must be charged before first use. Newer NiMH batteries are ready to be used when purchased, and have only 15% discharge in a year.

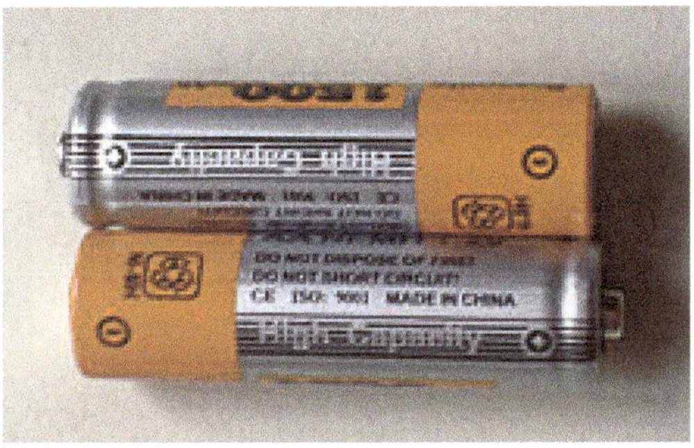

Rechargeable batteries

Some deterioration occurs on each charge–discharge cycle. Degradation usually occurs because electrolyte migrates away from the electrodes or because active material detaches from the electrodes. Low-capacity NiMH batteries (1,700–2,000 mA·h) can be charged some 1,000 times, whereas high-capacity NiMH batteries (above 2,500 mA·h) last about 500 cycles. NiCd batteries tend to be rated for 1,000 cycles before their internal resistance permanently increases beyond usable values.

Charge/discharge Speed

Fast charging increases component changes, shortening battery lifespan.

Overcharging

If a charger cannot detect when the battery is fully charged then overcharging is likely, damaging it.

Memory Effect

NiCd cells, if used in a particular repetitive manner, may show a decrease in capacity called "memory effect". The effect can be avoided with simple practices. NiMH cells, although similar in chemistry, suffer less from memory effect.

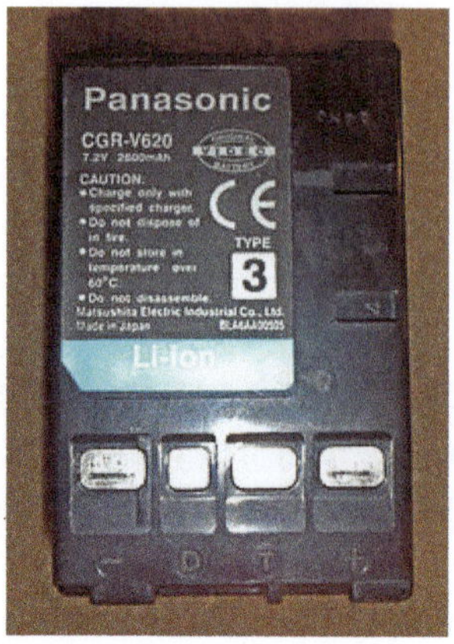

An analog camcorder [lithium ion] battery

Environmental Conditions

Automotive lead–acid rechargeable batteries must endure stress due to vibration, shock, and temperature range. Because of these stresses and sulfation of their lead plates, few automotive batteries last beyond six years of regular use. Automotive starting (SLI: *Starting, Lighting, Ignition*) batteries have many thin plates to maximize current. In general, the thicker the plates the longer the life. They are typically discharged only slightly before recharge.

"Deep-cycle" lead–acid batteries such as those used in electric golf carts have much thicker plates to extend longevity. The main benefit of the lead–acid battery is its low cost; its main drawbacks are large size and weight for a given capacity and voltage. Lead–acid batteries should never be discharged to below 20% of their capacity, because internal resistance will cause heat and damage when they are recharged. Deep-cycle lead–acid systems often use a low-charge warning light or a low-charge power cut-off switch to prevent the type of damage that will shorten the battery's life.

Storage

Battery life can be extended by storing the batteries at a low temperature, as in a refrigerator or freezer, which slows the side reactions. Such storage can extend the life of alkaline batteries by about 5%; rechargeable batteries can hold their charge much longer, depending upon type. To reach their maximum voltage, batteries must be returned to room temperature; discharging an alkaline battery at 250 mA at 0 °C is only half as efficient as at 20 °C. Alkaline battery manufacturers such as Duracell do not recommend refrigerating batteries.

Battery Sizes

Primary batteries readily available to consumers range from tiny button cells used for electric watches, to the No. 6 cell used for signal circuits or other long duration applications. Secondary cells are made in very large sizes; very large batteries can power a submarine or stabilize an electrical grid and help level out peak loads.

Hazards

Explosion

A battery explosion is caused by misuse or malfunction, such as attempting to recharge a primary (non-rechargeable) battery, or a short circuit. Car batteries are most likely to explode when a short-circuit generates very large currents. Car batteries produce hydrogen, which is very explosive, when they are overcharged (because of electrolysis of the water in the electrolyte). The amount of overcharging is usually very small and generates little hydrogen, which dissipates quickly. However, when "jumping" a car battery, the high current can cause the rapid release of large volumes of hydrogen, which can be ignited explosively by a nearby spark, for example, when disconnecting a jumper cable. When a battery is recharged at an excessive rate, an explosive gas mixture of hydrogen and oxygen may be produced faster than it can escape from within the battery, leading to pressure build-up and eventual bursting of the battery case. In extreme cases, battery acid may spray violently from the casing and cause injury. Overcharging—that is, attempting to charge a battery beyond its electrical capacity—can also lead to a battery explosion, in addition to leakage or irreversible damage. It may also cause damage to the charger or device in which the overcharged battery is later used. In addition, disposing of a battery via incineration may cause an explosion as steam builds up within the sealed case.

Leakage

Leak-damaged alkaline battery

Many battery chemicals are corrosive, poisonous or both. If leakage occurs, either spontaneously or through accident, the chemicals released may be dangerous. For example, disposable batteries often use a zinc "can" both as a reactant and as the container to hold the other reagents. If this kind of battery is over-discharged, the reagents can emerge through the cardboard and plastic that form the remainder of the container. The active chemical leakage can then damage or disable the equipment that the batteries power. For this reason, many electronic device manufacturers recommend removing the batteries from devices that will not be used for extended periods of time.

Toxic Materials

Many types of batteries employ toxic materials such as lead, mercury, and cadmium as an electrode or electrolyte. When each battery reaches end of life it must be disposed of to prevent environmental damage. Batteries are one form of electronic waste (e-waste). E-waste recycling services recover toxic substances, which can then be used for new batteries. Of the nearly three billion batteries purchased annually in the United States, about 179,000 tons end up in landfills across the country. In the United States, the Mercury-Containing and Rechargeable Battery Management Act of 1996 banned the sale of mercury-containing batteries, enacted uniform labeling requirements for rechargeable batteries and required that rechargeable batteries be easily removable. California and New York City prohibit the disposal of rechargeable batteries in solid waste, and along with Maine require recycling of cell phones. The rechargeable battery industry operates nationwide recycling programs in the United States and Canada, with dropoff points at local retailers.

The Battery Directive of the European Union has similar requirements, in addition to requiring increased recycling of batteries and promoting research on improved battery recycling methods. In accordance with this directive all batteries to be sold within the EU must be marked with the "collection symbol" (a crossed-out wheeled bin). This must cover at least 3% of the surface of prismatic batteries and 1.5% of the surface of cylindrical batteries. All packaging must be marked likewise.

Ingestion

Batteries may be harmful or fatal if swallowed. Small button cells can be swallowed, in particular by young children. While in the digestive tract, the battery's electrical discharge may lead to tissue damage; such damage is occasionally serious and can lead to death. Ingested disk batteries do not usually cause problems unless they become lodged in the gastrointestinal tract. The most common place for disk batteries to become lodged is the esophagus, resulting in clinical sequelae. Batteries that successfully traverse the esophagus are unlikely to lodge elsewhere. The likelihood that a disk battery will lodge in the esophagus is a function of the patient's age and battery size. Disk batteries of 16 mm have become lodged in the esophagi of 2 children younger than 1 year. Older children do not have problems with batteries smaller than 21–23 mm. Liquefaction necrosis may occur because sodium hydroxide is generated by the current produced by the battery (usually at the anode). Perforation has occurred as rapidly as 6 hours after ingestion.

Chemistry

Primary Batteries and their Characteristics

Secondary (Rechargeable) Batteries and their Characteristics

Chemistry	Cell voltage	Specific energy (MJ/kg)	Comments
NiCd	1.2	0.14	Inexpensive. High-/low-drain, moderate energy density. Can withstand very high discharge rates with virtually no loss of capacity. Moderate rate of self-discharge. Environmental hazard due to Cadmium – use now virtually prohibited in Europe.
Lead–acid	2.1	0.14	Moderately expensive. Moderate energy density. Moderate rate of self-discharge. Higher discharge rates result in considerable loss of capacity. Environmental hazard due to Lead. Common use – Automobile batteries
NiMH	1.2	0.36	Inexpensive. Performs better than alkaline batteries in higher drain devices. Traditional chemistry has high energy density, but also a high rate of self-discharge. Newer chemistry has low self-discharge rate, but also a ~25% lower energy density. Used in some cars.
NiZn	1.6	0.36	Moderately inexpensive. High drain device suitable. Low self-discharge rate. Voltage closer to alkaline primary cells than other secondary cells. No toxic components. Newly introduced to the market (2009). Has not yet established a track record. Limited size availability.
AgZn	1.86 1.5	0.46	Smaller volume than equivalent Li-ion. Extremely expensive due to silver. Very high energy density. Very high drain capable. For many years considered obsolete due to high silver prices. Cell suffers from oxidation if unused. Reactions are not fully understood. Terminal voltage very stable but suddenly drops to 1.5 volts at 70–80% charge (believed to be due to presence of both argentous and argentic oxide in positive plate – one is consumed first). Has been used in lieu of primary battery (moon buggy). Is being developed once again as a replacement for Li-ion.

Lithium ion	3.6	0.46	Very expensive. Very high energy density. Not usually available in "common" battery sizes. Very common in laptop computers, moderate to high-end digital cameras, camcorders, and cellphones. Very low rate of self-discharge. Tends to require either user awareness or a dedicated management system to slow down the gradual loss of capacity. Terminal voltage unstable (varies from 4.2 to 3.0 volts during discharge). Volatile: Chance of explosion if short-circuited, allowed to overheat, or not manufactured with rigorous quality standards.

Homemade Cells

Almost any liquid or moist object that has enough ions to be electrically conductive can serve as the electrolyte for a cell. As a novelty or science demonstration, it is possible to insert two electrodes made of different metals into a lemon, potato, etc. and generate small amounts of electricity. "Two-potato clocks" are also widely available in hobby and toy stores; they consist of a pair of cells, each consisting of a potato (lemon, et cetera) with two electrodes inserted into it, wired in series to form a battery with enough voltage to power a digital clock. Homemade cells of this kind are of no practical use.

A voltaic pile can be made from two coins (such as a nickel and a penny) and a piece of paper towel dipped in salt water. Such a pile generates a very low voltage but, when many are stacked in series, they can replace normal batteries for a short time.

Sony has developed a biological battery that generates electricity from sugar in a way that is similar to the processes observed in living organisms. The battery generates electricity through the use of enzymes that break down carbohydrates.

Lead acid cells can easily be manufactured at home, but a tedious charge/discharge cycle is needed to 'form' the plates. This is a process in which lead sulfate forms on the plates, and during charge is converted to lead dioxide (positive plate) and pure lead (negative plate). Repeating this process results in a microscopically rough surface, increasing the surface area, increasing the current the cell can deliver.

Daniell cells are easy to make at home. Aluminium–air batteries can be produced with high-purity aluminium. Aluminium foil batteries will produce some electricity, but are not efficient, in part because a significant amount of (combustible) hydrogen gas is produced.

Salt Bridge

A salt bridge, in electrochemistry, is a laboratory device used to connect the oxidation and reduction half-cells of a galvanic cell (voltaic cell), a type of electrochemical cell.

It maintains electrical neutrality within the internal circuit, preventing the cell from rapidly running its reaction to equilibrium. If no salt bridge were present, the solution in one half cell would accumulate negative charge and the solution in the other half cell would accumulate positive charge as the reaction proceeded, quickly preventing further reaction, and hence production of electricity.

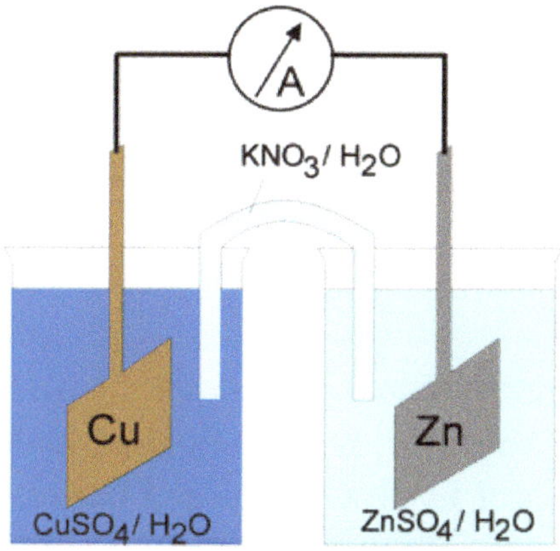

An electrochemical cell (resembling a Daniell cell) with a filter paper salt bridge. The paper has been soaked with a KNO_3 solution.

Salt bridges usually come in two types: glass tube and filter paper.

Glass Tube Bridges

One type of salt bridge consists of a U-shaped glass tube filled with a relatively inert electrolyte; usually potassium chloride or sodium chloride is used, although the diagram here illustrates the use of a potassium nitrate solution. The electrolyte is so chosen that

1. it does not react with any of the chemicals used in the cell
2. the anion and cation have similar conductivity, and hence similar migratory speed.

The electrolyte is often gelified with agar-agar to help prevent the intermixing of fluids which might otherwise occur.

The conductivity of a glass tube bridge depends mostly on the concentration of the electrolyte solution. At concentrations below saturation, an increase in concentration increases conductivity. Beyond-saturation electrolyte content and narrow tube diameter may both lower conductivity.

Filter Paper Bridges

The other type of salt bridge consists of a filter paper, also soaked with a relatively inert electrolyte, usually potassium chloride or sodium chloride because they are chemically inert. No gelification agent is required as the filter paper provides a solid medium for conduction.

Conductivity of this kind of salt bridge depends on a number of factors: the concentration of the electrolyte solution, the texture of the filter paper and the absorbing ability of the filter paper. Generally, smoother texture and higher absorbency equates to higher conductivity.

A porous disk or other porous barrier between the two half-cells may be used instead of a salt bridge; however, they basically serve the same purpose.

Fuel Cell

A fuel cell is a device that converts the chemical energy from a fuel into electricity through a chemical reaction of positively charged hydrogen ions with oxygen or another oxidizing agent. Fuel cells are different from batteries in requiring a continuous source of fuel and oxygen or air to sustain the chemical reaction, whereas in a battery the chemicals present in the battery react with each other to generate an electromotive force (emf). Fuel cells can produce electricity continuously for as long as these inputs are supplied.

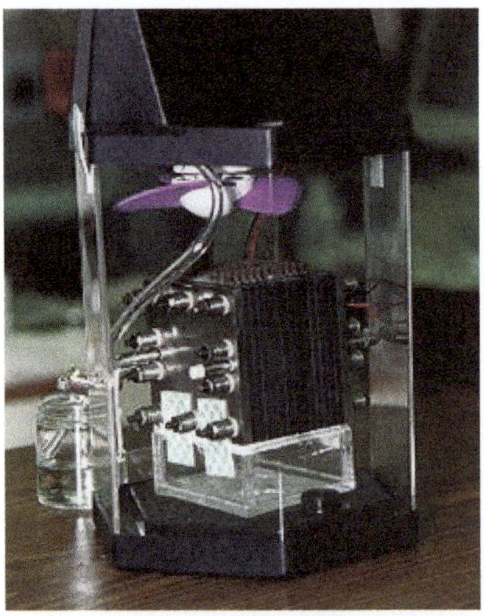

Demonstration model of a direct-methanol fuel cell. The actual fuel cell stack is the layered cube shape in the center of the image

The first fuel cells were invented in 1838. The first commercial use of fuel cells came more than a century later in NASA space programs to generate power for satellites and space capsules. Since then, fuel cells have been used in many other applications. Fuel cells are used for primary and backup power for commercial, industrial and residential buildings and in remote or inaccessible areas. They are also used to power fuel cell vehicles, including forklifts, automobiles, buses, boats, motorcycles and submarines.

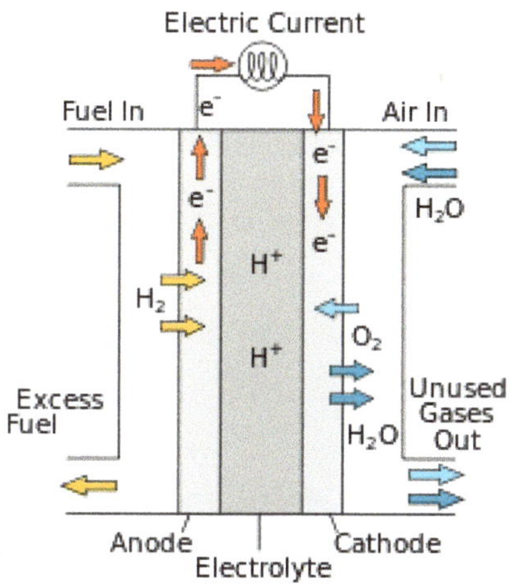

Scheme of a proton-conducting fuel cell

There are many types of fuel cells, but they all consist of an anode, a cathode, and an electrolyte that allows positively charged hydrogen ions (or protons) to move between the two sides of the fuel cell. The anode and cathode contain catalysts that cause the fuel to undergo oxidation reactions that generate positively charged hydrogen ions and electrons. The hydrogen ions are drawn through the electrolyte after the reaction. At the same time, electrons are drawn from the anode to the cathode through an external circuit, producing direct current electricity. At the cathode, hydrogen ions, electrons, and oxygen react to form water. As the main difference among fuel cell types is the electrolyte, fuel cells are classified by the type of electrolyte they use and by the difference in startup time ranging from 1 second for proton exchange membrane fuel cells (PEM fuel cells, or PEMFC) to 10 minutes for solid oxide fuel cells (SOFC). Individual fuel cells produce relatively small electrical potentials, about 0.7 volts, so cells are "stacked", or placed in series, to create sufficient voltage to meet an application's requirements. In addition to electricity, fuel cells produce water, heat and, depending on the fuel source, very small amounts of nitrogen dioxide and other emissions. The energy efficiency of a fuel cell is generally between 40–60%, or up to 85% efficient in cogeneration if waste heat is captured for use.

The fuel cell market is growing, and in 2013 Pike Research estimated that the stationary fuel cell market will reach 50 GW by 2020.

History

The first references to hydrogen fuel cells appeared in 1838. In a letter dated October 1838 but published in the December 1838 edition of *The London and Edinburgh Philosophical Magazine and Journal of Science*, Welsh physicist and barrister William Grove wrote about the development of his first crude fuel cells. He used a combination of sheet iron, copper and porcelain plates, and a solution of sulphate of copper and dilute acid. In a letter to the same publication written in December 1838 but published in June 1839,

German physicist Christian Friedrich Schönbein discussed the first crude fuel cell that he had invented. His letter discussed current generated from hydrogen and oxygen dissolved in water. Grove later sketched his design, in 1842, in the same journal. The fuel cell he made used similar materials to today's phosphoric-acid fuel cell. 9.

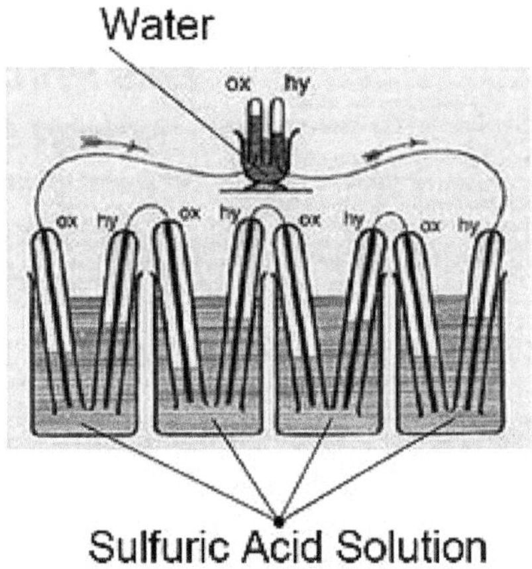

Sketch of William Grove's 1839 fuel cell

In 1939, British engineer Francis Thomas Bacon successfully developed a 5 kW stationary fuel cell. In 1955, W. Thomas Grubb, a chemist working for the General Electric Company (GE), further modified the original fuel cell design by using a sulphonated polystyrene ion-exchange membrane as the electrolyte. Three years later another GE chemist, Leonard Niedrach, devised a way of depositing platinum onto the membrane, which served as catalyst for the necessary hydrogen oxidation and oxygen reduction reactions. This became known as the "Grubb-Niedrach fuel cell". GE went on to develop this technology with NASA and McDonnell Aircraft, leading to its use during Project Gemini. This was the first commercial use of a fuel cell. In 1959, a team led by Harry Ihrig built a 15 kW fuel cell tractor for Allis-Chalmers, which was demonstrated across the U.S. at state fairs. This system used potassium hydroxide as the electrolyte and compressed hydrogen and oxygen as the reactants. Later in 1959, Bacon and his colleagues demonstrated a practical five-kilowatt unit capable of powering a welding machine. In the 1960s, Pratt and Whitney licensed Bacon's U.S. patents for use in the U.S. space program to supply electricity and drinking water (hydrogen and oxygen being readily available from the spacecraft tanks). In 1991, the first hydrogen fuel cell automobile was developed by Roger Billings.

UTC Power was the first company to manufacture and commercialize a large, stationary fuel cell system for use as a co-generation power plant in hospitals, universities and large office buildings.

In recognition of the fuel cell industry and America's role in fuel cell development, the US Senate recognized October 8, 2015 as National Hydrogen and Fuel Cell Day, passing S. RES 217. The date was chosen in recognition of the atomic weight of hydrogen (1.008).

Types of Fuel Cells; Design

Fuel cells come in many varieties; however, they all work in the same general manner. They are made up of three adjacent segments: the anode, the electrolyte, and the cathode. Two chemical reactions occur at the interfaces of the three different segments. The net result of the two reactions is that fuel is consumed, water or carbon dioxide is created, and an electric current is created, which can be used to power electrical devices, normally referred to as the load.

At the anode a catalyst oxidizes the fuel, usually hydrogen, turning the fuel into a positively charged ion and a negatively charged electron. The electrolyte is a substance specifically designed so ions can pass through it, but the electrons cannot. The freed electrons travel through a wire creating the electric current. The ions travel through the electrolyte to the cathode. Once reaching the cathode, the ions are reunited with the electrons and the two react with a third chemical, usually oxygen, to create water or carbon dioxide.

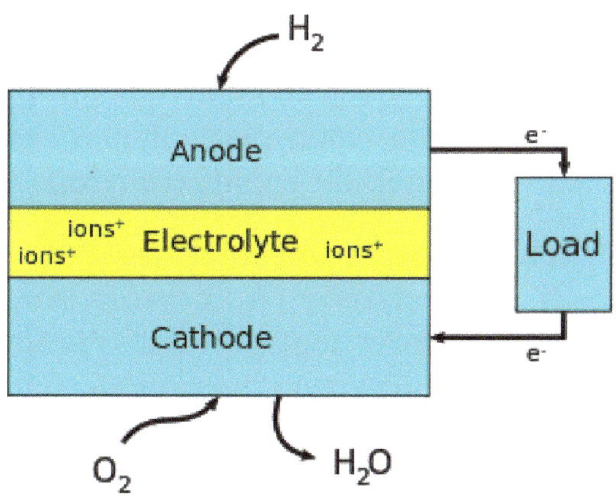

A block diagram of a fuel cell

The most important design features in a fuel cell are:

- The electrolyte substance. The electrolyte substance usually defines the *type* of fuel cell.
- The fuel that is used. The most common fuel is hydrogen.
- The anode catalyst breaks down the fuel into electrons and ions. The anode catalyst is usually made up of very fine platinum powder.
- The cathode catalyst turns the ions into the waste chemicals like water or carbon dioxide. The cathode catalyst is often made up of nickel but it can also be a nanomaterial-based catalyst.

A typical fuel cell produces a voltage from 0.6 V to 0.7 V at full rated load. Voltage decreases as current increases, due to several factors:

- Activation loss
- Ohmic loss (voltage drop due to resistance of the cell components and interconnections)
- Mass transport loss (depletion of reactants at catalyst sites under high loads, causing rapid loss of voltage).

To deliver the desired amount of energy, the fuel cells can be combined in series to yield higher voltage, and in parallel to allow a higher current to be supplied. Such a design is called a *fuel cell stack*. The cell surface area can also be increased, to allow higher current from each cell. Within the stack, reactant gases must be distributed uniformly over each of the cells to maximize the power output.

Proton Exchange Membrane Fuel Cells (PEMFCs)

In the archetypical hydrogen–oxide proton exchange membrane fuel cell design, a proton-conducting polymer membrane (typically nafion) contains the electrolyte solution that separates the anode and cathode sides. This was called a "solid polymer electrolyte fuel cell" (SPEFC) in the early 1970s, before the proton exchange mechanism was well-understood. (Notice that the synonyms "polymer electrolyte membrane" and "proton exchange mechanism" result in the same acronym.)

On the anode side, hydrogen diffuses to the anode catalyst where it later dissociates into protons and electrons. These protons often react with oxidants causing them to become what are commonly referred to as multi-facilitated proton membranes. The protons are conducted through the membrane to the cathode, but the electrons are forced to travel in an external circuit (supplying power) because the membrane is electrically insulating. On the cathode catalyst, oxygen molecules react with the electrons (which have traveled through the external circuit) and protons to form water.

In addition to this pure hydrogen type, there are hydrocarbon fuels for fuel cells, including diesel, methanol and chemical hydrides. The waste products with these types of fuel are carbon dioxide and water. When hydrogen is used, the CO2 is released when methane from natural gas is combined with steam, in a process called steam methane reforming, to produce the hydrogen. This can take place in a different location to the fuel cell, potentially allowing the hydrogen fuel cell to be used indoors—for example, in fork lifts.

Condensation of water produced by a PEMFC on the air channel wall. The gold wire around the cell ensures the collection of electric current.

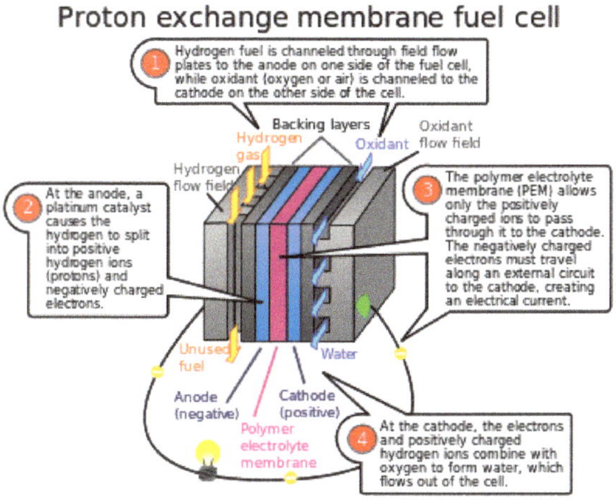

Construction of a high-temperature PEMFC: Bipolar plate as electrode with in-milled gas channel structure, fabricated from conductive composites (enhanced with graphite, carbon black, carbon fiber, and/or carbon nanotubes for more conductivity); Porous carbon papers; reactive layer, usually on the polymer membrane applied; polymer membrane.

The different components of a PEMFC are

1. bipolar plates,
2. electrodes,
3. catalyst,
4. membrane, and
5. the necessary hardware such as current collectors and gaskets.

The materials used for different parts of the fuel cells differ by type. The bipolar plates may be made of different types of materials, such as, metal, coated metal, graphite, flexible graphite, C–C composite, carbon–polymer composites etc. The membrane electrode assembly (MEA) is referred as the heart of the PEMFC and is usually made of a proton exchange membrane sandwiched between two catalyst-coated carbon papers. Platinum and/or similar type of noble metals are usually used as the catalyst for PEMFC. The electrolyte could be a polymer membrane.

Proton Exchange Membrane Fuel Cell Design Issues

- Cost. In 2013, the Department of Energy estimated that 80-kW automotive fuel cell system costs of US$67 per kilowatt could be achieved, assuming volume production of 100,000 automotive units per year and US$55 per kilowatt could be achieved, assuming volume production of 500,000 units per year. Many companies are working on techniques to reduce cost in a variety of ways including reducing the amount of platinum needed in each individual cell. Ballard Power Systems has experimented with a catalyst enhanced with carbon silk, which allows a 30% reduction (1 mg/cm² to 0.7 mg/cm²) in platinum usage without reduction in performance. Monash University, Melbourne uses PEDOT as a cathode. A 2011 published study doi: 10.1021/ja1112904 documented the first metal-free electrocatalyst using relatively inexpensive doped carbon nanotubes, which are less than 1% the cost of platinum and are of equal or superior performance. A recently published article demonstrated how the environmental burdens change when using carbon nanotubes as carbon substrate for platinum.

- Water and air management (in PEMFCs). In this type of fuel cell, the membrane must be hydrated, requiring water to be evaporated at precisely the same rate that it is produced. If water is evaporated too quickly, the membrane dries, resistance across it increases, and eventually it will crack, creating a gas "short circuit" where hydrogen and oxygen combine directly, generating heat that will damage the fuel cell. If the water is evaporated too slowly, the electrodes will flood, preventing the reactants from reaching the catalyst and stopping the reaction. Methods to manage water in cells are being developed like electroosmotic pumps focusing on flow control. Just as in a combustion engine, a steady ratio between the reactant and oxygen is necessary to keep the fuel cell operating efficiently.

- Temperature management. The same temperature must be maintained throughout the cell in order to prevent destruction of the cell through thermal loading. This is particularly challenging as the $2H_2 + O_2 \rightarrow 2H_2O$ reaction is highly exothermic, so a large quantity of heat is generated within the fuel cell.

- Durability, service life, and special requirements for some type of cells. Stationary fuel cell applications typically require more than 40,000 hours of reliable operation at a temperature of −35 °C to 40 °C (−31 °F to 104 °F), while automotive fuel cells require a 5,000-hour lifespan (the equivalent of 240,000 km (150,000 mi)) under extreme temperatures. Current service life is 2,500 hours (about 75,000 miles). Automotive engines must also be able to start reliably at −30 °C (−22 °F) and have a high power-to-volume ratio (typically 2.5 kW per liter).

- Limited carbon monoxide tolerance of some (non-PEDOT) cathodes.

Phosphoric Acid Fuel Cell (PAFC)

Phosphoric acid fuel cells (PAFC) were first designed and introduced in 1961 by G. V. Elmore and H. A. Tanner. In these cells phosphoric acid is used as a non-conductive electrolyte to pass positive hydrogen ions from the anode to the cathode. These cells commonly work in temperatures of 150 to 200 degrees Celsius. This high temperature will cause heat and energy loss if the heat is not removed and used properly. This heat can be used to produce steam for air conditioning systems or any other thermal energy consuming system. Using this heat in cogeneration can enhance the efficiency of phosphoric acid fuel cells from 40–50% to about 80%. Phosphoric acid, the electrolyte used in PAFCs, is a non-conductive liquid acid which forces electrons to travel from anode to cathode through an external electrical circuit. Since the hydrogen ion production rate on the anode is small, platinum is used as catalyst to increase this ionization rate. A key disadvantage of these cells is the use of an acidic electrolyte. This increases the corrosion or oxidation of components exposed to phosphoric acid.

Solid Acid Fuel Cell (SAFC)

Solid acid fuel cells (SAFCs) are characterized by the use of a solid acid material as the electrolyte. At low temperatures, solid acids have an ordered molecular structure like most salts. At warmer temperatures (between 140 to 150 degrees Celsius for $CsHSO_4$), some solid acids undergo a phase transition to become highly disordered "superprotonic" structures, which increases conductivity by several orders of magnitude. The first proof-of-concept SAFCs were developed in 2000 using cesium hydrogen sulfate ($CsHSO_4$). Current SAFC systems use cesium dihydrogen phosphate (CsH_2PO_4) and have demonstrated lifetimes in the thousands of hours.

High-temperature Fuel Cells

SOFC

Solid oxide fuel cells (SOFCs) use a solid material, most commonly a ceramic material called yttria-stabilized zirconia (YSZ), as the electrolyte. Because SOFCs are made entirely of solid materials, they are not limited to the flat plane configuration of other types of fuel cells and are often designed as rolled tubes. They require high operating temperatures (800–1000 °C) and can be run on a variety of fuels including natural gas.

SOFCs are unique since in those, negatively charged oxygen ions travel from the cathode (positive side of the fuel cell) to the anode (negative side of the fuel cell) instead of positively charged hydrogen ions travelling from the anode to the cathode, as is the case in all other types of fuel cells. Oxygen gas is fed through the cathode, where it absorbs electrons to create oxygen ions. The oxygen ions then travel through the electrolyte to react with hydrogen gas at the anode. The reaction at the anode produces electricity and water as by-products. Carbon dioxide may also be a by-product depending on the fuel, but the carbon emissions from an SOFC system are less than those from a fossil fuel combustion plant. The chemical reactions for the SOFC system can be expressed as follows:

Anode Reaction: $2H_2 + 2O^{2-} \rightarrow 2H_2O + 4e^-$

Cathode Reaction: $O_2 + 4e^- \rightarrow 2O^{2-}$

Overall Cell Reaction: $2H_2 + O_2 \rightarrow 2H_2O$

SOFC systems can run on fuels other than pure hydrogen gas. However, since hydrogen is necessary for the reactions listed above, the fuel selected must contain hydrogen atoms. For the fuel cell to operate, the fuel must be converted into pure hydrogen gas. SOFCs are capable of internally reforming light hydrocarbons such as methane (natural gas), propane and butane. These fuel cells are at an early stage of development.

Challenges exist in SOFC systems due to their high operating temperatures. One such challenge is the potential for carbon dust to build up on the anode, which slows down the internal reforming process. Research to address this "carbon coking" issue at the University of Pennsylvania has shown that the use of copper-based cermet (heat-resistant materials made of ceramic and metal) can reduce coking and the loss of performance. Another disadvantage of SOFC systems is slow start-up time, making SOFCs less useful for mobile applications. Despite these disadvantages, a high operating temperature provides an advantage by removing the need for a precious metal catalyst like platinum, thereby reducing cost. Additionally, waste heat from SOFC systems may be captured and reused, increasing the theoretical overall efficiency to as high as 80%–85%.

The high operating temperature is largely due to the physical properties of the YSZ electrolyte. As temperature decreases, so does the ionic conductivity of YSZ. Therefore, to obtain optimum performance of the fuel cell, a high operating temperature is required. According to their website, Ceres Power, a UK SOFC fuel cell manufacturer, has developed a method of reducing the operating temperature of their SOFC system to 500–600 degrees Celsius. They replaced the commonly used YSZ electrolyte with a CGO (cerium gadolinium oxide) electrolyte. The lower operating temperature allows them to use stainless steel instead of ceramic as the cell substrate, which reduces cost and start-up time of the system.

Hydrogen-oxygen Fuel Cell

The hydrogen-oxygen fuel cell or alkaline fuel cell was designed and first demonstrated publicly by Francis Thomas Bacon in 1959. It was used as a primary source of electrical energy in the Apollo space program. The cell consists of two porous carbon electrodes impregnated with a suitable catalyst such as Pt, Ag, CoO, etc. The space between the two electrodes is filled with a concentrated solution of KOH or NaOH which serves as an electrolyte. H_2 gas and O_2 gas are bubbled into the electrolyte through the porous carbon electrodes. Thus the overall reaction involves the combination of hydrogen gas and oxygen gas to form water. The cell runs continuously until the reactant's supply is exhausted. This type of cell operates efficiently in the temperature range 343 K to 413 K and provides a potential of about 0.9 V.

MCFC

Molten carbonate fuel cells (MCFCs) require a high operating temperature, 650 °C (1,200 °F), similar to SOFCs. MCFCs use lithium potassium carbonate salt as an electrolyte, and this salt liquefies at high temperatures, allowing for the movement of charge within the cell – in this case, negative carbonate ions.

Like SOFCs, MCFCs are capable of converting fossil fuel to a hydrogen-rich gas in the anode, eliminating the need to produce hydrogen externally. The reforming process creates CO_2 emissions. MCFC-compatible fuels include natural gas, biogas and gas produced from coal. The hydrogen in the gas reacts with carbonate ions from the electrolyte to produce water, carbon dioxide, electrons and small amounts of other chemicals. The electrons travel through an external circuit creating electricity and return to the cathode. There, oxygen from the air and carbon dioxide recycled from the anode react with the electrons to form carbonate ions that replenish the electrolyte, completing the circuit. The chemical reactions for an MCFC system can be expressed as follows:

Anode Reaction: $CO_3^{2-} + H_2 \rightarrow H_2O + CO_2 + 2e^-$

Cathode Reaction: $CO_2 + \frac{1}{2}O_2 + 2e^- \rightarrow CO_3^{2-}$

Overall Cell Reaction: $H_2 + \frac{1}{2}O_2 \rightarrow H_2O$

As with SOFCs, MCFC disadvantages include slow start-up times because of their high operating temperature. This makes MCFC systems not suitable for mobile applications, and this technology will most likely be used for stationary fuel cell purposes. The main challenge of MCFC technology is the cells' short life span. The high-temperature and carbonate electrolyte lead to corrosion of the anode and cathode. These factors accelerate the degradation of MCFC components, decreasing the durability and cell life. Researchers are addressing this problem by exploring corrosion-resistant materials for components as well as fuel cell designs that may increase cell life without decreasing performance.

MCFCs hold several advantages over other fuel cell technologies, including their resistance to impurities. They are not prone to "carbon coking", which refers to carbon build-up on the anode that results in reduced performance by slowing down the internal fuel reforming process. Therefore, carbon-rich fuels like gases made from coal are compatible with the system. The Department of Energy claims that coal, itself, might even be a fuel option in the future, assuming the system can be made resistant to impurities such as sulfur and particulates that result from converting coal into hydrogen. MCFCs also have relatively high efficiencies. They can reach a fuel-to-electricity efficiency of 50%, considerably higher than the 37–42% efficiency of a phosphoric acid fuel cell plant. Efficiencies can be as high as 65% when the fuel cell is paired with a turbine, and 85% if heat is captured and used in a Combined Heat and Power (CHP) system.

FuelCell Energy, a Connecticut-based fuel cell manufacturer, develops and sells MCFC fuel cells. The company says that their MCFC products range from 300 kW to 2.8 MW systems that achieve 47% electrical efficiency and can utilize CHP technology to obtain higher overall efficiencies. One product, the DFC-ERG, is combined with a gas turbine and, according to the company, it achieves an electrical efficiency of 65%.

Comparison of Fuel Cell Types

Secondary (rechargeable) batteries and their characteristics	Electrolyte	Qualified power (W)	Working temperature (°C)	Efficiency (cell)	Efficiency (system)	Status	Cost (USD/W)
Metal hydride fuel cell	Aqueous alkaline solution		> -20 (50% P_{peak} @ 0 °C)			Commercial / Research	
Electro-galvanic fuel cell	Aqueous alkaline solution		< 40			Commercial / Research	
Direct formic acid fuel cell (DFAFC)	Polymer membrane (ionomer)	< 50 W	< 40			Commercial / Research	
Zinc-air battery	Aqueous alkaline solution		< 40			Mass production	
Microbial fuel cell	Polymer membrane or humic acid		< 40			Research	
Upflow microbial fuel cell (UMFC)			< 40			Research	
Regenerative fuel cell	Polymer membrane (ionomer)		< 50			Commercial / Research	
Direct borohydride fuel cell	Aqueous alkaline solution		70			Commercial	
Alkaline fuel cell	Aqueous alkaline solution	10 – 100 kW	< 80	60–70%	62%	Commercial / Research	
Direct methanol fuel cell	Polymer membrane (ionomer)	100 mW – 1 kW	90–120	20–30%	10–25%	Commercial / Research	125
Reformed methanol fuel cell	Polymer membrane (ionomer)	5 W – 100 kW	250–300 (Reformer) 125–200 (PBI)	50–60%	25–40%	Commercial / Research	

Direct-ethanol fuel cell	Polymer membrane (ionomer)	< 140 mW/cm²	> 25 ? 90–120			Research	
Proton exchange membrane fuel cell	Polymer membrane (ionomer)	1 W – 500 kW	50–100 (Nafion) 120–200 (PBI)	50–70%	30–50%	Commercial / Research	50–100
RFC – Redox	Liquid electrolytes with redox shuttle and polymer membrane (Ionomer)	1 kW – 10 MW				Research	
Phosphoric acid fuel cell	Molten phosphoric acid (H_3PO_4)	< 10 MW	150-200	55%	40% Co-Gen: 90%	Commercial / Research	4–4.50
Solid acid fuel cell	H^+-conducting oxyanion salt (solid acid)	10 W - 1 kW	200-300	55-60%	40-45%	Commercial / Research	
Molten carbonate fuel cell	Molten alkaline carbonate	100 MW	600–650	55%	45-55%	Commercial / Research	
Tubular solid oxide fuel cell (TSOFC)	O^{2-}-conducting ceramic oxide	< 100 MW	850–1100	60–65%	55–60%	Commercial / Research	
Protonic ceramic fuel cell	H^+-conducting ceramic oxide		700			Research	
Direct carbon fuel cell	Several different		700–850	80%	70%	Commercial / Research	
Planar Solid oxide fuel cell	O^{2-}-conducting ceramic oxide	< 100 MW	500–1100	60–65%	55–60%	Commercial / Research	
Enzymatic Biofuel Cells	Any that will not denature the enzyme		< 40			Research	
Magnesium-Air Fuel Cell	Salt water		−20 to 55	90%		Commercial / Research	

Efficiency of Leading Fuel Cell Types

Glossary of Terms in table:

- Anode: The electrode at which oxidation (a loss of electrons) takes place. For fuel cells and other galvanic cells, the anode is the negative terminal; for electrolytic cells (where electrolysis occurs), the anode is the positive terminal.
- Aqueous solution: a: of, relating to, or resembling water b : made from, with, or by water.
- Catalyst: A chemical substance that increases the rate of a reaction without being consumed; after the reaction, it can potentially be recovered from the reaction mixture and is chemically unchanged. The catalyst lowers the activation energy required, allowing the reaction to proceed more quickly or at a lower temperature. In a fuel cell, the catalyst facilitates the reaction of oxygen and hydrogen. It is usually made of platinum powder very thinly coated onto carbon paper or cloth. The catalyst is rough and porous so the maximum surface area of the platinum can be exposed to the hydrogen or oxygen. The platinum-coated side of the catalyst faces the membrane in the fuel cell.
- Cathode: The electrode at which reduction (a gain of electrons) occurs. For fuel cells and other galvanic cells, the cathode is the positive terminal; for electrolytic cells (where electrolysis occurs), the cathode is the negative terminal.
- Electrolyte: A substance that conducts charged ions from one electrode to the other in a fuel cell, battery, or electrolyzer.
- Fuel Cell Stack: Individual fuel cells connected in a series. Fuel cells are stacked to increase voltage.
- Matrix: something within or from which something else originates, develops, or takes form.
- Membrane: The separating layer in a fuel cell that acts as electrolyte (an ion-exchanger) as well as a barrier film separating the gases in the anode and cathode compartments of the fuel cell.
- Molten Carbonate Fuel Cell (MCFC): A type of fuel cell that contains a molten carbonate electrolyte. Carbonate ions (CO_3^{2-}) are transported from the cathode to the anode. Operating temperatures are typically near 650 °C.
- Phosphoric acid fuel cell (PAFC): A type of fuel cell in which the electrolyte consists of concentrated phosphoric acid (H_3PO_4). Protons (H+) are transported from the anode to the cathode. The operating temperature range is generally 160–220 °C.

- Polymer Electrolyte Membrane (PEM): A fuel cell incorporating a solid polymer membrane used as its electrolyte. Protons (H+) are transported from the anode to the cathode. The operating temperature range is generally 60–100 °C.
- Solid Oxide Fuel Cell (SOFC): A type of fuel cell in which the electrolyte is a solid, nonporous metal oxide, typically zirconium oxide (ZrO_2) treated with Y_2O_3, and O^{2-} is transported from the cathode to the anode. Any CO in the reformate gas is oxidized to CO_2 at the anode. Temperatures of operation are typically 800–1,000 °C.
- Solution: a: an act or the process by which a solid, liquid, or gaseous substance is homogeneously mixed with a liquid or sometimes a gas or solid, b : a homogeneous mixture formed by this process; especially : a single-phase liquid system, c : the condition of being dissolved

Theoretical Maximum Efficiency

The energy efficiency of a system or device that converts energy is measured by the ratio of the amount of useful energy put out by the system ("output energy") to the total amount of energy that is put in ("input energy") or by useful output energy as a percentage of the total input energy. In the case of fuel cells, useful output energy is measured in electrical energy produced by the system. Input energy is the energy stored in the fuel. According to the U.S. Department of Energy, fuel cells are generally between 40–60% energy efficient. This is higher than some other systems for energy generation. For example, the typical internal combustion engine of a car is about 25% energy efficient. In combined heat and power (CHP) systems, the heat produced by the fuel cell is captured and put to use, increasing the efficiency of the system to up to 85–90%.

The theoretical maximum efficiency of any type of power generation system is never reached in practice, and it does not consider other steps in power generation, such as production, transportation and storage of fuel and conversion of the electricity into mechanical power. However, this calculation allows the comparison of different types of power generation. The maximum theoretical energy efficiency of a fuel cell is 83%, operating at low power density and using pure hydrogen and oxygen as reactants (assuming no heat recapture) According to the World Energy Council, this compares with a maximum theoretical efficiency of 58% for internal combustion engines. While these efficiencies are not approached in most real world applications, high-temperature fuel cells (solid oxide fuel cells or molten carbonate fuel cells) can theoretically be combined with gas turbines to allow stationary fuel cells to come closer to the theoretical limit. A gas turbine would capture heat from the fuel cell and turn it into mechanical energy to increase the fuel cell's operational efficiency. This solution has been predicted to increase total efficiency to as much as 80%.

In Practice

In a fuel-cell vehicle the tank-to-wheel efficiency is greater than 45% at low loads and shows average values of about 36% when a driving cycle like the NEDC (New European

Driving Cycle) is used as test procedure. The comparable NEDC value for a Diesel vehicle is 22%. In 2008 Honda released a demonstration fuel cell electric vehicle (the Honda FCX Clarity) with fuel stack claiming a 60% tank-to-wheel efficiency.

It is also important to take losses due to fuel production, transportation, and storage into account. Fuel cell vehicles running on compressed hydrogen may have a power-plant-to-wheel efficiency of 22% if the hydrogen is stored as high-pressure gas, and 17% if it is stored as liquid hydrogen. Fuel cells cannot store energy like a battery, except as hydrogen, but in some applications, such as stand-alone power plants based on discontinuous sources such as solar or wind power, they are combined with electrolyzers and storage systems to form an energy storage system. Most hydrogen, however, is produced by steam methane reforming, and so most hydrogen production emits carbon dioxide. The overall efficiency (electricity to hydrogen and back to electricity) of such plants (known as *round-trip efficiency*), using pure hydrogen and pure oxygen can be "from 35 up to 50 percent", depending on gas density and other conditions. While a much cheaper lead–acid battery might return about 90%, the electrolyzer/fuel cell system can store indefinite quantities of hydrogen, and is therefore better suited for long-term storage.

Solid-oxide fuel cells produce heat from the recombination of the oxygen and hydrogen. The ceramic can run as hot as 800 degrees Celsius. This heat can be captured and used to heat water in a micro combined heat and power (m-CHP) application. When the heat is captured, total efficiency can reach 80–90% at the unit, but does not consider production and distribution losses. CHP units are being developed today for the European home market.

Professor Jeremy P. Meyers, in the Electrochemical Society journal *Interface* in 2008, wrote, "While fuel cells are efficient relative to combustion engines, they are not as efficient as batteries, due primarily to the inefficiency of the oxygen reduction reaction (and ... the oxygen evolution reaction, should the hydrogen be formed by electrolysis of water).... [T]hey make the most sense for operation disconnected from the grid, or when fuel can be provided continuously. For applications that require frequent and relatively rapid start-ups ... where zero emissions are a requirement, as in enclosed spaces such as warehouses, and where hydrogen is considered an acceptable reactant, a [PEM fuel cell] is becoming an increasingly attractive choice [if exchanging batteries is inconvenient]". In 2013 military organisations are evaluating fuel cells to significantly reduce the battery weight carried by soldiers.

Applications

Type 212 submarine with fuel cell propulsion of the German Navy in dry dock

Power

Stationary fuel cells are used for commercial, industrial and residential primary and backup power generation. Fuel cells are very useful as power sources in remote locations, such as spacecraft, remote weather stations, large parks, communications centers, rural locations including research stations, and in certain military applications. A fuel cell system running on hydrogen can be compact and lightweight, and have no major moving parts. Because fuel cells have no moving parts and do not involve combustion, in ideal conditions they can achieve up to 99.9999% reliability. This equates to less than one minute of downtime in a six-year period.

Since fuel cell electrolyzer systems do not store fuel in themselves, but rather rely on external storage units, they can be successfully applied in large-scale energy storage, rural areas being one example. There are many different types of stationary fuel cells so efficiencies vary, but most are between 40% and 60% energy efficient. However, when the fuel cell's waste heat is used to heat a building in a cogeneration system this efficiency can increase to 85%. This is significantly more efficient than traditional coal power plants, which are only about one third energy efficient. Assuming production at scale, fuel cells could save 20–40% on energy costs when used in cogeneration systems. Fuel cells are also much cleaner than traditional power generation; a fuel cell power plant using natural gas as a hydrogen source would create less than one ounce of pollution (other than CO_2) for every 1,000 kW·h produced, compared to 25 pounds of pollutants generated by conventional combustion systems. Fuel Cells also produce 97% less nitrogen oxide emissions than conventional coal-fired power plants.

One such pilot program is operating on Stuart Island in Washington State. There the Stuart Island Energy Initiative has built a complete, closed-loop system: Solar panels power an electrolyzer, which makes hydrogen. The hydrogen is stored in a 500-U.S.-gallon (1,900 L) tank at 200 pounds per square inch (1,400 kPa), and runs a ReliOn fuel cell to provide full electric back-up to the off-the-grid residence. Another closed system loop was unveiled in late 2011 in Hempstead, NY.

Fuel cells can be used with low-quality gas from landfills or waste-water treatment plants to generate power and lower methane emissions. A 2.8 MW fuel cell plant in California is said to be the largest of the type.

Cogeneration

Combined heat and power (CHP) fuel cell systems, including Micro combined heat and power (MicroCHP) systems are used to generate both electricity and heat for homes, office building and factories. The system generates constant electric power (selling excess power back to the grid when it is not consumed), and at the same time produces hot air and water from the waste heat. As the result CHP systems have the potential to save primary energy as they can make use of waste heat which is generally rejected by thermal energy conversion systems. A typical capacity range of home fuel cell is 1–3 kW_{el} / 4–8 kW_{th}. CHP systems linked to absorption chillers use their waste heat for refrigeration.

The waste heat from fuel cells can be diverted during the summer directly into the ground providing further cooling while the waste heat during winter can be pumped directly into the building. The University of Minnesota owns the patent rights to this type of system

Co-generation systems can reach 85% efficiency (40–60% electric + remainder as thermal). Phosphoric-acid fuel cells (PAFC) comprise the largest segment of existing CHP products worldwide and can provide combined efficiencies close to 90%. Molten Carbonate (MCFC) and Solid Oxide Fuel Cells (SOFC) are also used for combined heat and power generation and have electrical energy efficiences around 60%. Disadvantages of co-generation systems include slow ramping up and down rates, high cost and short lifetime. Also their need to have a hot water storage tank to smooth out the thermal heat production was a serious disadvantage in the domestic market place where space in domestic properties is at a great premium.

Delta-ee consultants stated in 2013 that with 64% of global sales the fuel cell micro-combined heat and power passed the conventional systems in sales in 2012. The Japanese ENE FARM project will pass 100,000 FC mCHP systems in 2014, 34.213 PEMFC and 2.224 SOFC were installed in the period 2012-2014, 30,000 units on LNG and 6,000 on LPG.

Fuel Cell Electric Vehicles (FCEVs)

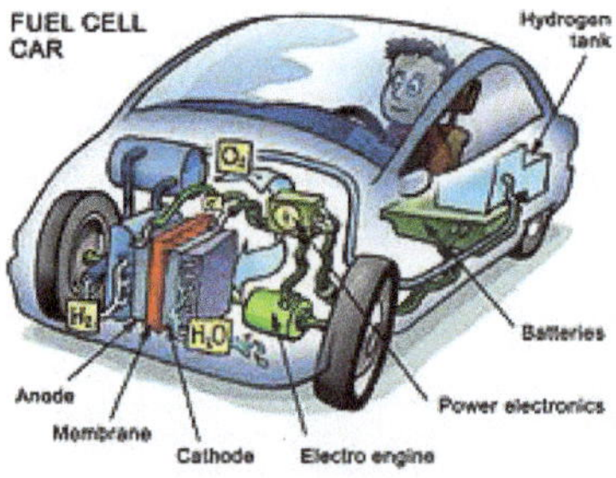

Configuration of components in a fuel cell car

Toyota Mirai

Automobiles

As of 2015, two fuel cell vehicles have been introduced for commercial lease and sale in limited quantities: the Toyota Mirai and the Hyundai ix35 FCEV. Additional demonstration models include the Honda FCX Clarity, and Mercedes-Benz F-Cell. As of June 2011 demonstration FCEVs had driven more than 4,800,000 km (3,000,000 mi), with more than 27,000 refuelings. Demonstration fuel cell vehicles have been produced with "a driving range of more than 400 km (250 mi) between refueling". They can be refueled in less than 5 minutes. The U.S. Department of Energy's Fuel Cell Technology Program claims that, as of 2011, fuel cells achieved 53–59% efficiency at one-quarter power and 42–53% vehicle efficiency at full power, and a durability of over 120,000 km (75,000 mi) with less than 10% degradation. In a Well-to-Wheels simulation analysis that "did not address the economics and market constraints", General Motors and its partners estimated that per mile traveled, a fuel cell electric vehicle running on compressed gaseous hydrogen produced from natural gas could use about 40% less energy and emit 45% less greenhouse gasses than an internal combustion vehicle. A lead engineer from the Department of Energy whose team is testing fuel cell cars said in 2011 that the potential appeal is that "these are full-function vehicles with no limitations on range or refueling rate so they are a direct replacement for any vehicle. For instance, if you drive a full sized SUV and pull a boat up into the mountains, you can do that with this technology and you can't with current battery-only vehicles, which are more geared toward city driving."

In 2014, Toyota introduced its first fuel cell vehicle in Japan, the Mirai, at a price of less than US$100,000, although former European Parliament President Pat Cox estimates that Toyota will initially lose about $100,000 on each Mirai sold. Hyundai introduced the limited production Hyundai ix35 FCEV. Other manufacturers that have announced intentions to sell fuel cell electric vehicles commercially by 2016 include General Motors, Honda, Mercedes-Benz, and Nissan.

Criticism

Some experts believe that hydrogen fuel cell cars will never become economically competitive with other technologies or that it will take decades for them to become profitable. Elon Musk stated in 2015 that fuel cells for use in cars will never be commercially viable because of the inefficiency of producing, transporting and storing hydrogen and the flammability of the gas, among other reasons. Professor Jeremy P. Meyers estimated in 2008 that cost reductions over a production ramp-up period will take about 20 years after fuel-cell cars are introduced before they will be able to compete commercially with current market technologies, including gasoline internal combustion engines. In 2011, the chairman and CEO of General Motors, Daniel Akerson, stated that while the cost of hydrogen fuel cell cars is decreasing: "The car is still too expensive and probably won't be practical until the 2020-plus period, I don't know."

In 2012, Lux Research, Inc. issued a report that stated: "The dream of a hydrogen economy ... is no nearer". It concluded that "Capital cost ... will limit adoption to a mere 5.9 GW" by 2030, providing "a nearly insurmountable barrier to adoption, except in niche applications". The analysis concluded that, by 2030, PEM stationary market will reach $1 billion, while the vehicle market, including forklifts, will reach a total of $2 billion. Other analyses cite the lack of an extensive hydrogen infrastructure in the U.S. as an ongoing challenge to Fuel Cell Electric Vehicle commercialization. In 2006, a study for the IEEE showed that for hydrogen produced via electrolysis of water: "Only about 25% of the power generated from wind, water, or sun is converted to practical use." The study further noted that "Electricity obtained from hydrogen fuel cells appears to be four times as expensive as electricity drawn from the electrical transmission grid. ... Because of the high energy losses [hydrogen] cannot compete with electricity." Furthermore, the study found: "Natural gas reforming is not a sustainable solution". "The large amount of energy required to isolate hydrogen from natural compounds (water, natural gas, biomass), package the light gas by compression or liquefaction, transfer the energy carrier to the user, plus the energy lost when it is converted to useful electricity with fuel cells, leaves around 25% for practical use."

Joseph Romm, the author of *The Hype About Hydrogen* (2005), devoted two articles in 2014 to updating his critique of the use of fuel cells in cars. He stated that FCVs still had not overcome the following issues: high cost of the vehicles, high fueling cost, and a lack of fuel-delivery infrastructure. "It would take several miracles to overcome all of those problems simultaneously in the coming decades." Most importantly, he said, "FCVs aren't green" because of escaping methane during natural gas extraction and when hydrogen is produced, as 95% of it is, using the steam reforming process. He concluded that renewable energy cannot economically be used to make hydrogen for an FCV fleet "either now or in the future." Greentech Media's analyst reached similar conclusions in 2014. In 2015, *Clean Technica* listed some of the disadvantages of hydrogen fuel cell vehicles. So did *Car Throttle*. Another *Clean Technica* writer concluded, "while hydrogen may have a part to play in the world of energy storage (especially seasonal storage), it looks like a dead end when it comes to mainstream vehicles."

Buses

As of August 2011, there were a total of approximately 100 fuel cell buses deployed around the world. Most buses are produced by UTC Power, Toyota, Ballard, Hydrogenics, and Proton Motor. UTC buses had accumulated over 970,000 km (600,000 mi) of driving by 2011. Fuel cell buses have a 39–141% higher fuel economy than diesel buses and natural gas buses. Fuel cell buses have been deployed around the world including in Whistler, Canada; San Francisco, United States; Hamburg, Germany; Shanghai, China; London, England; and São Paulo, Brazil.

Toyota FCHV-BUS at the Expo 2005.

The Fuel Cell Bus Club is a global cooperative effort in trial fuel cell buses. Notable projects include:

- 12 fuel cell buses are being deployed in the Oakland and San Francisco Bay area of California.
- Daimler AG, with 36 experimental buses powered by Ballard Power Systems fuel cells, completed a successful three-year trial in eleven cities, in January 2007.
- A fleet of Thor buses with UTC Power fuel cells was deployed in California, operated by SunLine Transit Agency.

The first Brazilian hydrogen fuel cell bus prototype in Brazil was deployed in São Paulo. The bus was manufactured in Caxias do Sul and the hydrogen fuel will be produced in São Bernardo do Campo from water through electrolysis. The program, called "*Ônibus Brasileiro a Hidrogênio*" (Brazilian Hydrogen Autobus), includes three additional buses.

Forklifts

A fuel cell forklift (also called a fuel cell lift truck) is a fuel cell-powered industrial forklift truck used to lift and transport materials. In 2013 there were over 4,000 fuel cell forklifts used in material handling in the US, of which only 500 received funding from DOE (2012). The global market is 1 million fork lifts per year. Fuel cell fleets are operated by various companies, including Sysco Foods, FedEx Freight, GENCO (at Wegmans, Coca-Cola, Kimberly Clark, and Whole Foods), and H-E-B Grocers. Europe demonstrated 30 fuel cell forklifts with Hylift and extended it with HyLIFT-EUROPE to 200 units, with other projects in France and Austria. Pike Research stated in 2011 that fuel cell-powered forklifts will be the largest driver of hydrogen fuel demand by 2020.

Most companies in Europe and the US do not use petroleum-powered forklifts, as these vehicles work indoors where emissions must be controlled and instead use electric forklifts. Fuel cell-powered forklifts can provide benefits over battery-powered forklifts as they can work for a full 8-hour shift on a single tank of hydrogen and can be refueled in 3 minutes. Fuel cell-powered forklifts can be used in refrigerated warehouses, as their performance is not degraded by lower temperatures. The FC units are often designed as drop-in replacements.

Motorcycles and Bicycles

In 2005 a British manufacturer of hydrogen-powered fuel cells, Intelligent Energy (IE), produced the first working hydrogen run motorcycle called the ENV (Emission Neutral Vehicle). The motorcycle holds enough fuel to run for four hours, and to travel 160 km (100 mi) in an urban area, at a top speed of 80 km/h (50 mph). In 2004 Honda developed a fuel-cell motorcycle that utilized the Honda FC Stack.

Other examples of motorbikes and bicycles that use hydrogen fuel cells include the Taiwanese company APFCT's scooter using the fueling system from Italy's Acta SpA and the Suzuki Burgman scooter with an IE fuel cell that received EU Whole Vehicle Type Approval in 2011. Suzuki Motor Corp. and IE have announced a joint venture to accelerate the commercialization of zero-emission vehicles.

Airplanes

Boeing researchers and industry partners throughout Europe conducted experimental flight tests in February 2008 of a manned airplane powered only by a fuel cell and lightweight batteries. The fuel cell demonstrator airplane, as it was called, used a proton exchange membrane (PEM) fuel cell/lithium-ion battery hybrid system to power an electric motor, which was coupled to a conventional propeller. In 2003, the world's first propeller-driven airplane to be powered entirely by a fuel cell was flown. The fuel cell was a stack design that allowed the fuel cell to be integrated with the plane's aerodynamic surfaces.

Fuel cell-powered unmanned aerial vehicles (UAV) include a Horizon fuel cell UAV that set the record distance flown for a small UAV in 2007. The military is interested in this application because of its low noise, low thermal signature and ability to attain high altitude. In 2009 the Naval Research Laboratory's (NRL's) Ion Tiger utilized a hydrogen-powered fuel cell and flew for 23 hours and 17 minutes. Fuel cells are also in use to provide auxiliary power in aircraft, replacing fossil fuel generators that were previously used to start the engines and power on board electrical needs.

In January 2016 a Raptor E1 drone made a successful test flight using a fuel cell that was lighter than the lithium-ion battery it replaced. The flight lasted 10 minutes at an altitude of 80 metres (260 ft), although the fuel cell reportedly had enough fuel to fly for two hours. The fuel was contained in approximately 100 solid 1 square centimetre (0.16 sq in) pellets composed of a proprietary chemical within an unpressurized cartridge. The pellets are physically robust and operate at temperatures as warm as 50 °C (122 °F). The cell was from Arcola Energy.

Boats

The world's first fuel-cell boat HYDRA used an AFC system with 6.5 kW net output. Iceland has committed to converting its vast fishing fleet to use fuel cells to provide auxiliary power by 2015 and, eventually, to provide primary power in its boats. Amsterdam recently introduced its first fuel cell-powered boat that ferries people around the city's canals.

The world's first certified fuel cell boat (HYDRA), in Leipzig/Germany

Submarines

The Type 212 submarines of the German and Italian navies use fuel cells to remain submerged for weeks without the need to surface.

The U212A is a non-nuclear submarine developed by German naval shipyard Howaldtswerke Deutsche Werft. The system consists of nine PEM fuel cells, providing between 30 kW and 50 kW each. The ship is silent, giving it an advantage in the detection of other submarines. A naval paper has theorized about the possibility of a nuclear-fuel cell hybrid whereby the fuel cell is used when silent operations are required and then replenished from the Nuclear reactor (and water).

Portable Power Systems

Portable power systems that use fuel cells can be used in the leisure sector (i.e. RVs, cabins, marine), the industrial sector (i.e. power for remote locations including gas/oil wellsites, communication towers, security, weather stations), and in the military sector. SFC Energy is a German manufacturer of direct methanol fuel cells for a variety of portable power systems. Ensol Systems Inc. is an integrator of portable power systems, using the SFC Energy DMFC.

Other Applications

- Providing power for base stations or cell sites
- Distributed generation
- Emergency power systems are a type of fuel cell system, which may include lighting, generators and other apparatus, to provide backup resources in a crisis

or when regular systems fail. They find uses in a wide variety of settings from residential homes to hospitals, scientific laboratories, data centers,

- telecommunication equipment and modern naval ships.
- An uninterrupted power supply (UPS) provides emergency power and, depending on the topology, provide line regulation as well to connected equipment by supplying power from a separate source when utility power is not available. Unlike a standby generator, it can provide instant protection from a momentary power interruption.
- Base load power plants
- Solar Hydrogen Fuel Cell Water Heating
- Hybrid vehicles, pairing the fuel cell with either an ICE or a battery.
- Notebook computers for applications where AC charging may not be readily available.
- Portable charging docks for small electronics (e.g. a belt clip that charges a cell phone or PDA).
- Smartphones, laptops and tablets.
- Small heating appliances
- Food preservation, achieved by exhausting the oxygen and automatically maintaining oxygen exhaustion in a shipping container, containing, for example, fresh fish.
- Breathalyzers, where the amount of voltage generated by a fuel cell is used to determine the concentration of fuel (alcohol) in the sample.
- Carbon monoxide detector, electrochemical sensor.

Fueling Stations

There were over 85 hydrogen refueling stations in the U.S. in 2010. Some of these have closed, and In November 2013, *The New York Times* reported that there were "10 hydrogen stations available to the public in the entire United States: one in Columbia, S.C., eight in Southern California and the one in Emeryville". In 2013 the Department of Energy launched H2USA focused on advancing the hydrogen infrastructure. As of July 2015, there were 12 public hydrogen fueling stations in the US, 10 of which were in California.

The first public hydrogen refueling station in Iceland operated from 2003 to 2007. It served three buses in the public transport net of Reykjavík. The station produced its own hydrogen with an electrolyzing unit. The 14 stations in Germany were planned to be expanded to 50 by 2015 through its public private partnership Now GMBH.

Hydrogen fueling station.

Japan has a hydrogen highway, as part of the Japan hydrogen fuel cell project. Twelve hydrogen fueling stations had been built in 11 cities in Japan by 2012. Canada and Sweden also had planned hydrogen highways. Norway plans to build a network of hydrogen station between the major cities, starting in 2017.

Markets and Economics

In 2012, fuel cell industry revenues exceeded $1 billion market value worldwide, with Asian pacific countries shipping more than 3/4 of the fuel cell systems worldwide. However, as of January 2014, no public company in the industry had yet become profitable. There were 140,000 fuel cell stacks shipped globally in 2010, up from 11,000 shipments in 2007, and from 2011 to 2012 worldwide fuel cell shipments had an annual growth rate of 85%. Tanaka Kikinzoku expanded its manufacturing facilities in 2011. Approximately 50% of fuel cell shipments in 2010 were stationary fuel cells, up from about a third in 2009, and the four dominant producers in the Fuel Cell Industry were the United States, Germany, Japan and South Korea. The Department of Energy Solid State Energy Conversion Alliance found that, as of January 2011, stationary fuel cells generated power at approximately $724 to $775 per kilowatt installed. In 2011, Bloom Energy, a major fuel cell supplier, said that its fuel cells generated power at 9–11 cents per kilowatt-hour, including the price of fuel, maintenance, and hardware.

Industry groups predict that there are sufficient platinum resources for future demand, and in 2007, research at Brookhaven National Laboratory suggested that platinum could be replaced by a gold-palladium coating, which may be less susceptible to poisoning and thereby improve fuel cell lifetime. Another method would use iron and sulphur instead of platinum. This would lower the cost of a fuel cell (as the platinum in a regular fuel cell costs around US$1,500, and the same amount of iron costs only around US$1.50). The concept was being developed by a coalition of the John Innes Centre and the University of Milan-Bicocca. PEDOT cathodes are immune to monoxide poisoning.

In 2016, Samsung "decided to drop fuel cell-related business projects, as the outlook of the market isn't good".

Research and Development

- 2005: Georgia Institute of Technology researchers used triazole to raise the operating temperature of PEM fuel cells from below 100 °C to over 125 °C, claiming this will require less carbon-monoxide purification of the hydrogen fuel.
- 2008: Monash University, Melbourne used PEDOT as a cathode.
- 2009: Researchers at the University of Dayton, in Ohio, showed that arrays of vertically grown carbon nanotubes could be used as the catalyst in fuel cells. The same year, a nickel bisdiphosphine-based catalyst for fuel cells was demonstrated.
- 2013: British firm ACAL Energy developed a fuel cell that it said can run for 10,000 hours in simulated driving conditions. It asserted that the cost of fuel cell construction can be reduced to $40/kW (roughly $9,000 for 300 HP).
- 2014: Researchers in Imperial College London developed a new method for regeneration of hydrogen sulfide contaminated PEFCs. They recovered 95–100% of the original performance of a hydrogen sulfide contaminated PEFC. They were successful in rejuvenating a SO_2 contaminated PEFC too. This regeneration method is applicable to multiple cell stacks.

Galvanometer

A galvanometer is an electromechanical instrument for detecting and measuring electric current. The most common use of galvanometers was as analog measuring instruments, called ammeters, used to measure the direct current (flow of electric charge) through an electric circuit. A galvanometer works as an actuator, by producing a rotary deflection (of a "pointer"), in response to electric current flowing through a coil in a constant magnetic field.

An early D'Arsonval galvanometer showing magnet and rotating coil.

Galvanometers developed from the observation that the needle of a magnetic compass is deflected near a wire that has electric current flowing through it, first described by Hans Oersted in 1820. They were the first instruments used to detect and measure small amounts of electric currents. The name comes from the Italian electricity researcher Luigi Galvani, who in 1791 discovered the principle of the frog galvanoscope – that electric current would make the legs of a dead frog jerk.

Sensitive galvanometers have been essential for the development of science and technology in many fields. For example, they enabled long range communication through submarine cables, such as the earliest Transatlantic telegraph cables, and were essential to discovering the electrical activity of the heart and brain, by their fine measurements of current.

Galvanometers also had widespread use as the visualising part in other kinds of analog meters, for example in light meters, VU meters, etc., where they were used to measure and display the output of other sensors. Today the main type of galvanometer mechanism, still in use, is the moving coil, D'Arsonval/Weston type.

Operation

Modern galvanometers, of the D'Arsonval/Weston type, are constructed with a small pivoting coil of wire in the field of a permanent magnet. The coil is attached to a thin pointer that traverses a calibrated scale. A tiny torsion spring pulls the coil and pointer to the zero position.

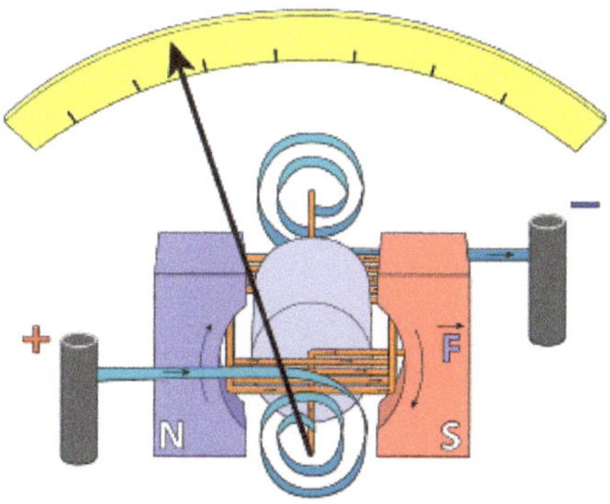

Diagram of D'Arsonval/Weston type galvanometer. As the current flows from + through the coil (the orange part) to −, a magnetic field is generated in the coil. This field is counteracted by the permanent magnet and forces the coil to twist, moving the pointer, in relation to the field's strength caused by the flow of current.

When a direct current (DC) flows through the coil, the coil generates a magnetic field. This field acts against the permanent magnet. The coil twists, pushing against the spring, and moves the pointer. The hand points at a scale indicating the electric current. Careful design of the pole pieces ensures that the magnetic field is uniform, so that the angular deflection of the pointer is proportional to the current. A useful meter generally contains provision for damping the mechanical resonance of the moving coil and pointer, so that the pointer settles quickly to its position without oscillation.

The basic sensitivity of a meter might be, for instance, 100 microamperes full scale (with a voltage drop of, say, 50 millivolts at full current). Such meters are often calibrated to read some other quantity that can be converted to a current of that magnitude. The use of current dividers, often called shunts, allows a meter to be calibrated to measure larger currents. A meter can be calibrated as a DC voltmeter if the resistance of the coil is known by calculating the voltage required to generate a full scale current. A meter can be configured to read other voltages by putting it in a voltage divider circuit. This is generally done by placing a resistor in series with the meter coil. A meter can be used to read resistance by placing it in series with a known voltage (a battery) and an adjustable resistor. In a preparatory step, the circuit is completed and the resistor adjusted to produce full scale deflection. When an unknown resistor is placed in series in the circuit the current will be less than full scale and an appropriately calibrated scale can display the value of the previously unknown resistor.

These capabilities to translate different kinds of electric quantities, in to pointer movements, make the galvanometer ideal for turning output of other sensors that outputs electricity (in some form or another), into something that can be read by a human.

Because the pointer of the meter is usually a small distance above the scale of the meter, parallax error can occur when the operator attempts to read the scale line that "lines up" with the pointer. To counter this, some meters include a mirror along the markings of the principal scale. The accuracy of the reading from a mirrored scale is improved by positioning one's head while reading the scale so that the pointer and the reflection of the pointer are aligned; at this point, the operator's eye must be directly above the pointer and any parallax error has been minimized.

Uses

Probably the largest use of galvanometers was of the D'Arsonval/Weston type used in analog meters in electronic equipment. Since the 1980s, galvanometer-type analog meter movements have been displaced by analog to digital converters (ADCs) for many uses. A digital panel meter (DPM) contains an analog to digital converter and numeric display. The advantages of a digital instrument are higher precision and accuracy, but factors such as power consumption or cost may still favour application of analog meter movements.

Modern closed-loop galvanometer-driven laser scanning mirror from Scanlab.

Modern Uses

Most modern uses for the galvanometer mechanism are in positioning and control systems. Galvanometer mechanisms are divided into moving magnet and moving coil galvanometers; in addition, they are divided into *closed-loop* and *open-loop* - or *resonant* - types.

Mirror galvanometer systems are used as beam positioning or beam steering elements in laser scanning systems. For example, for material processing with high-power lasers, closed loop mirror galvanometer mechanisms are used with servo control systems. These are typically high power galvanometers and the newest galvanometers designed for beam steering applications can have frequency responses over 10 kHz with appropriate servo technology. Closed-loop mirror galvanometers are also used in similar ways in stereolithography, laser sintering, laser engraving, laser beam welding, laser TVs, laser displays and in imaging applications such as retinal scanning with Optical Coherence Tomography (OCT). Almost all of these galvanometers are of the moving magnet type.

Open loop, or resonant mirror galvanometers, are mainly used in some types of laser-based bar-code scanners, printing machines, imaging applications, military applications and space systems. Their non-lubricated bearings are especially of interest in applications that require functioning in a high vacuum.

A galvanometer mechanism (center part), used in an automatic exposure unit of an 8 mm film camera, together with a photoresistor (seen in the hole on top of the leftpart).

Moving coil type galvanometer mechanisms (called 'voice coils' by hard disk manufacturers) are used for controlling the *head positioning* servos in hard disk drives and CD/DVD players, in order to keep mass (and thus access times), as low as possible.

Past Uses

A major early use for galvanometers was for finding faults in telecommunications cables. They were superseded in this application late in the 20th century by time-domain reflectometers.

Galvanometer mechanisms were also used to get readings from photoresistors in the metering mechanisms of film cameras (as seen in the image to the right).

In analog strip chart recorders such as used in electrocardiographs, electroencephalographs and polygraphs, galvanometer mechanisms were used to position the *pen*. Strip chart recorders with galvanometer driven pens may have a full scale frequency response of 100 Hz and several centimeters of deflection.

History

The deflection of a magnetic compass needle by current in a wire was first described by Hans Oersted in 1820. The phenomenon was studied both for its own sake and as a means of measuring electric current. The earliest galvanometer was reported by Johann Schweigger at the University of Halle on 16 September 1820. André-Marie Ampère also contributed to its development. Early designs increased the effect of the magnetic field generated by the current by using multiple turns of wire. The instruments were at first called "multipliers" due to this common design feature. The term "galvanometer," in common use by 1836, was derived from the surname of Italian electricity researcher Luigi Galvani, who in 1791 discovered that electric current would make a dead frog's leg jerk.

Thomson mirror galvanometer, patented in 1858.

Originally, the instruments relied on the Earth's magnetic field to provide the restoring force for the compass needle. These were called "tangent" galvanometers and had to be oriented before use. Later instruments of the "astatic" type used opposing magnets to become independent of the Earth's field and would operate in any orientation. The most sensitive form, the Thomson or mirror galvanometer, was patented in 1858 by William Thomson (Lord Kelvin) as an improvement of an earlier design invented in 1826 by Johann Christian Poggendorff. Thomson's design, was able to detect very rapid current changes, by using small magnets attached to a lightweight mirror, suspended by a thread, instead of a compass needle. The deflection of a light beam on the mirror greatly magnified the deflection induced by small currents. Alternatively, the deflection of the suspended magnets could be observed directly through a microscope.

The ability to measure quantitatively voltage and current allowed Georg Ohm to formulate that – the voltage across a conductor is directly proportional to the current through it – Ohm's Law.

The early moving-magnet form of galvanometer had the disadvantage that it was affected by any magnets or iron masses near it, and its deflection was not linearly proportional to the current. In 1882 Jacques-Arsène d'Arsonval and Marcel Deprez developed a form with a stationary permanent magnet and a moving coil of wire, suspended by fine wires which provided both an electrical connection to the coil and the restoring torque to return to the zero position. An iron tube between the magnet's pole pieces defined a circular gap through which the coil rotated. This gap produced a consistent, radial magnetic field across the coil, giving a linear response throughout the instrument's range. A mirror attached to the coil deflected a beam of light to indicate the coil position. The concentrated magnetic field and delicate suspension made these instruments sensitive; d'Arsonval's initial instrument could detect ten microamperes.

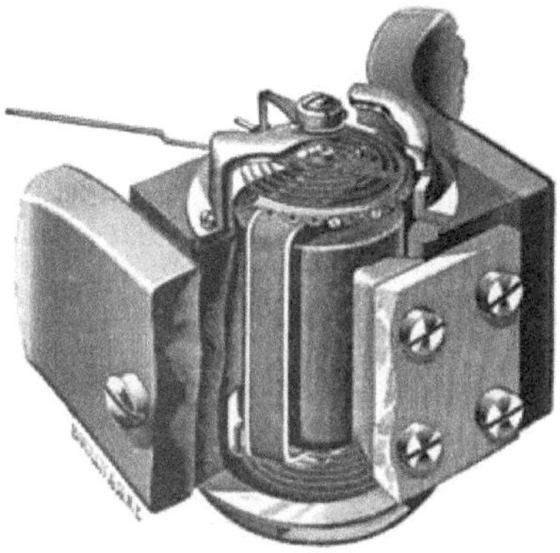

D'Arsonval/Weston galvanometer (ca. 1900). Part of the magnet's left pole piece is broken out to show the coil.

Edward Weston extensively improved the design. He replaced the fine wire suspension with a pivot, and provided restoring torque and electrical connections through spiral springs rather like those of a wristwatch balance wheel hairspring. He developed a method of stabilizing the magnetic field of the permanent magnet, so the instrument would have consistent accuracy over time. He replaced the light beam and mirror with a knife-edge pointer that could be read directly. A mirror under the pointer, in the same plane as the scale, eliminated parallax observation error. To maintain the field strength, Weston's design used a very narrow circumferential slot through which the coil moved, with a minimal air-gap. This improved linearity of pointer deflection with respect to coil current. Finally, the coil was wound on a light-weight form made of conductive metal, which acted as a damper. By 1888, Edward Weston had patented and brought out a commercial form of this instrument, which became a standard electrical equipment component. It was known as a "portable" instrument because it was affected very little by mounting position or by transporting it from place to place. This design is almost universally used in moving-coil meters today.

Initially laboratory instruments relying on the Earth's own magnetic field to provide restoring force for the pointer, galvanometers were developed into compact, rugged, sensitive portable instruments essential to the development of electro-technology.

Types

Some galvanometers use a solid pointer on a scale to show measurements; other very sensitive types use a miniature mirror and a beam of light to provide mechanical amplification of low-level signals.

Tangent Galvanometer

A tangent galvanometer is an early measuring instrument used for the measurement of electric current. It works by using a compass needle to compare a magnetic field generated by the unknown current to the magnetic field of the Earth. It gets its name from its operating principle, the tangent law of magnetism, which states that the tangent of the angle a compass needle makes is proportional to the ratio of the strengths of the two perpendicular magnetic fields. It was first described by Claude Pouillet in 1837.

Tangent galvanometer made by J. H. Bunnell Co. around 1890.

A tangent galvanometer consists of a coil of insulated copper wire wound on a circular non-magnetic frame. The frame is mounted vertically on a horizontal base provided with levelling screws. The coil can be rotated on a vertical axis passing through its centre. A compass box is mounted horizontally at the centre of a circular scale. It consists of a tiny, powerful magnetic needle pivoted at the centre of the coil. The magnetic needle is free to rotate in the horizontal plane. The circular scale is divided into four quadrants. Each quadrant is graduated from 0° to 90°. A long thin aluminium pointer is attached to the needle at its centre and at right angle to it. To avoid errors due to parallax, a plane mirror is mounted below the compass needle.

In operation, the instrument is first rotated until the magnetic field of the Earth, indicated by the compass needle, is parallel with the plane of the coil. Then the unknown current is applied to the coil. This creates a second magnetic field on the axis of the coil, perpendicular to the Earth's magnetic field. The compass needle responds to the vector sum of the two fields, and deflects to an angle equal to the tangent of the ratio of the two fields. From the angle read from the compass's scale, the current could be found from a table. The current supply wires have to be wound in a small helix, like a pig's tail, otherwise the field due to the wire will affect the compass needle and an incorrect reading will be obtained.

Theory

Top view of a tangent galvanometer made about 1950. The indicator needle of the compass is perpendicular to the shorter, black magnetic needle.

The galvanometer is oriented so that the plane of the coil is vertical and aligned along parallel to the horizontal component B_H of the Earth's magnetic field (i.e. parallel to the local "magnetic meridian"). When an electric current flows through the galvanometer coil, a second magnetic field B is created. At the center of the coil, where the compass needle is located, the coil's field is perpendicular to the plane of the coil. The magnitude of the coil's field is:

where I is the current in amperes, n is the number of turns of the coil and r is the radius of the coil. These two perpendicular magnetic fields add vectorially, and the compass needle points along the direction of their resultant B_H+B. The current in the coil causes the compass needle to rotate by an angle θ:

$$\theta = \tan^{-1} \frac{B}{B_H}$$

From tangent law, $B = B_H \tan \theta$, i.e.

$$\frac{\mu_0 n I}{2r} = B_H \tan \theta$$

or

$$I = \left(\frac{2 r B_H}{\mu_0 n} \right) \tan \theta$$

or $I = K \tan \theta$, where K is called the Reduction Factor of the tangent galvanometer.

One problem with the tangent galvanometer is that its resolution degrades at both high currents and low currents. The maximum resolution is obtained when the value of θ is 45°. When the value of θ is close to 0° or 90°, a large percentage change in the current will only move the needle a few degrees.

Geomagnetic Field Measurement

A tangent galvanometer can also be used to measure the magnitude of the horizontal component of the geomagnetic field. When used in this way, a low-voltage power source, such as a battery, is connected in series with a rheostat, the galvanometer, and an ammeter. The galvanometer is first aligned so that the coil is parallel to the geomagnetic field, whose direction is indicated by the compass when there is no current through the coils. The battery is then connected and the rheostat is adjusted until the compass needle deflects 45 degrees from the geomagnetic field, indicating that the magnitude of the magnetic field at the center of the coil is the same as that of the horizontal component of the geomagnetic field. This field strength can be calculated from the current as measured by the ammeter, the number of turns of the coil, and the radius of the coils.

Astatic Galvanometer

Unlike a compass-needle galvanometer, the astatic galvanometer has two magnetic needles parallel to each other, but with the magnetic poles reversed. The needle assembly is suspended by a silk thread, and has no net magnetic dipole moment. It is not affected by the earth's magnetic field. The lower needle is inside the current sensing coils and is deflected by the magnetic field created by the passing current.

Astatic galvanometer.

The *astatic galvanometer* was developed by Leopoldo Nobili in 1825.

Mirror galvanometer

To get a higher level of precision, for extremely sensitive measuring equipment, the mirror galvanometer substituted the pointer with a lightweight mirror. Thus a beam of light, reflected from the mirror, acted as a long mass-less pointer. This kind of galvanometer were for example used as the receivers in early trans-Atlantic telegraph systems. In a device called an oscillograph, the moving beam of light was used, to produce graphs of current versus time, by recording measurements on photographic film. The string galvanometer was a type of mirror galvanometer so sensitive that it was used to make the first electrocardiogram of the electrical activity of the human heart.

Ballistic Galvanometer

A ballistic galvanometer is a type of sensitive galvanometer for measuring the quantity of charge discharged through it. In reality it is an integrator, unlike a current-measuring galvanometer, the moving part has a large moment of inertia that gives it a long oscillation period. It can be either of the moving coil or moving magnet type, commonly it is a mirror galvanometer.

Photocathode

A photocathode is a negatively charged electrode in a light detection device such as a photomultiplier or phototube that is coated with a photosensitive compound. When this is struck by a quantum of light (photon), the absorbed energy causes electron emission due to the photoelectric effect.

Uses

For many years the photocathode was the only practical method for converting light to an electron current. As such it tends to function as a form of 'electric film' and shared many characteristics of photography. It was therefore the key element in opto-electronic devices, such as TV camera tubes like the orthocon and vidicon, and in image tubes such as intensifiers, converters, and dissectors. Simple phototubes were used for motion detectors and counters.

Phototubes have been used for years in movie projectors to read the sound tracks on the edge of movie film.

The more recent development of solid state optical devices such as photodiodes has reduced the use of photocathodes to cases where they still remain superior to semiconductor devices.

Construction

Photocathodes operate in a vacuum, so their design parallels vacuum tube technology. Since most cathodes are sensitive to air the construction of photocathodes typically occurs after the enclosure has been evacuated. In operation the photocathode requires an electric field with a nearby positive anode to assure electron emission.

Photocathodes divide into two broad groups; transmission and reflective. A transmission type is typically a coating upon a glass window where the light strikes one surface and electrons exit from the opposite surface. A reflective type is typically formed on an opaque metal electrode base, where the light enters and the electrons exit from the same side. A variation is the double reflection type, where the metal base is mirror-like, causing light that passed through the photocathode without causing emission to be bounced back for a second try. This mimics the retina on many mammals.

The effectiveness of a photocathode is commonly expressed as quantum efficiency, that being the ratio of emitted electrons vs. impinging quanta (of light). The efficiency varies with construction as well, as it can be improved with a stronger electric field.

Coatings

Although a plain metallic cathode will exhibit photoelectric properties, the specialized coating greatly increases the effect. A photocathode usually consists of alkali metals with very low work functions.

The coating releases electrons much more readily than the underlying metal, allowing it to detect the low-energy photons in infrared radiation. The lens transmits the radiation from the object being viewed to a layer of coated glass. The photons strike the metal surface and transfer electrons to its rear side. The freed electrons are then collected to produce the final image.

Photocathode Materials

- Ag-O-Cs, also called S-1. This was the first compound photocathode material, developed in 1929. Sensitivity from 300 nm to 1200 nm. Since Ag-O-Cs has a higher dark current than more modern materials photomultiplier tubes with this photocathode material are nowadays used only in the infrared region with cooling.

- Sb-Cs (antimony-caesium) has a spectral response from UV to visible and is mainly used in reflection-mode photocathodes.

- Bialkali (antimony-rubidium-caesium Sb-Rb-Cs, antimony-potassium-caesium Sb-K-Cs). Spectral response range similar to the Sb-Cs photocathode, but with higher sensitivity and lower dark current than Sb-Cs. They have sensitivity well matched to the most common scintillator materials and so are frequently used for ionizing radiation measurement in scintillation counters.

- High temperature bialkali or low noise bialkali (sodium-potassium-antimony, Na-K-Sb). This material is often used in oil well logging since it can withstand temperatures up to 175 °C. At room temperatures, this photocathode operates with very low dark current, making it ideal for use in photon counting applications.
- Multialkali (sodium-potassium-antimony-caesium, Na-K-Sb-Cs), also called S-20. The multialkali photocathode has a wide spectral response from the ultraviolet to near infrared region. It is widely used for broad-band spectrophotometers and photon counting applications. The long wavelength response can be extended to 930 nm by a special photocathode activation processing.
- GaAs (gallium(II) arsenide). This photocathode material covers a wider spectral response range than multialkali, from ultraviolet to 930 nm.
- InGaAs (indium gallium arsenide). Extended sensitivity in the infrared range compared to GaAs. Moreover, in the range between 900 nm and 1000 nm, InGaAs has a much better signal to noise ratio than Ag-O-Cs. With special manufacturing techniques this photocathode can operate up to 1700 nm.
- Cs-Te, Cs-I (caesium-telluride, caesium iodide). These materials are sensitive to vacuum UV and UV rays but not to visible light and are therefore referred to as solar blind. Cs-Te is insensitive to wavelengths longer than 320 nm, and Cs-I to those longer than 200 nm.

References

- Kipnis, Nahum (2003) "Changing a theory: the case of Volta's contact electricity", Nuova Voltiana, Vol. 5. Università degli studi di Pavia, 2003 ISBN 88-203-3273-6. pp. 144–146
- Clarke, Edwin; Jacyna, L. S. (1992) Nineteenth-Century Origins of Neuroscientific Concepts, University of California Press. ISBN 0-520-07879-9. p. 199
- Haughton, Brian (2007) Hidden History: Lost Civilizations, Secret Knowledge, and Ancient Mysteries. Career Press. ISBN 1564148971.
- Crompton, T. R. (2000-03-20). Battery Reference Book (third ed.). Newnes. p. Glossary 3. ISBN 0080499953. Retrieved 2016-03-18.
- Pauling, Linus (1988). "15: Oxidation-Reduction Reactions; Electrolysis.". General Chemistry. New York: Dover Publications, Inc. p. 539. ISBN 978-0-486-65622-9.
- Hogendoorn, Bob (2010). Heinemann Chemistry Enhanced (2). Melbourne, Australia: Pearson Australia. p. 416. ISBN 9781442537552.
- Joseph F. Keithley The story of electrical and magnetic measurements: from 500 B.C. to the 1940s, John Wiley and Sons, 1999 ISBN 0-7803-1193-0, pp. 196-198
- Coxworth, Ben (2016-02-08). "Drone flight powered by lightweight hydrogen-producing pellets". www.gizmag.com. Retrieved 2016-02-09.

- Kakati, Biraj Kumar; Unnikrishnan, Anusree; Rajalakshmi, Natarajan; Jafri, RI; Dhathathreyan, KS. "Kucernak". Anthony RJ. 41 (12): 5598–5604. doi:10.1016/j.ijhydene.2016.01.077. Retrieved 22 March 2016.
- Martin, Christopher (March 10, 2014). "Plug, FuelCell Climb as 'Experiments' Seen as Profitable". Bloomberg.com. Retrieved December 28, 2015.
- Kakati, BK. "In-situ O3 rejuvenation of SO2 contaminated Polymer Electrolyte Fuel Cell: Electrochemistry, single cell and 5-cells stack studies" (PDF). 5th European PEFC & H2 Forum. Retrieved 14 July 2015.
- White, Charlie (2008-07-31). "Hydrogen fuel cell vehicles are a fraud". DVICE. Archived from the original on 2014-06-19. Retrieved 2015-09-21.
- Williams, K.R. (1 February 1994). "Francis Thomas Bacon. 21 December 1904-24 May 1992" (PDF). Biographical Memoirs of Fellows of the Royal Society. 39: 2–9. doi:10.1098/rsbm.1994.0001. Retrieved January 5, 2015.

Various Reactions and Processes in Electrochemistry

In order to stop iron or steel from rusting a coat of zinc is applied on it. This process of applying zinc is known as galvanization. The process of bonding steel to protect it from corrosion is termed as electrogalvaning. This chapter helps the reader in understanding the various reactions and processes in electrochemistry.

Galvanic Corrosion

Galvanic corrosion (also called bimetallic corrosion or contact corrosion) is an electrochemical process in which one metal corrodes preferentially to another when both metals are in electrical contact, in the presence of an electrolyte. This same galvanic reaction is exploited in primary batteries to generate an electrical voltage.

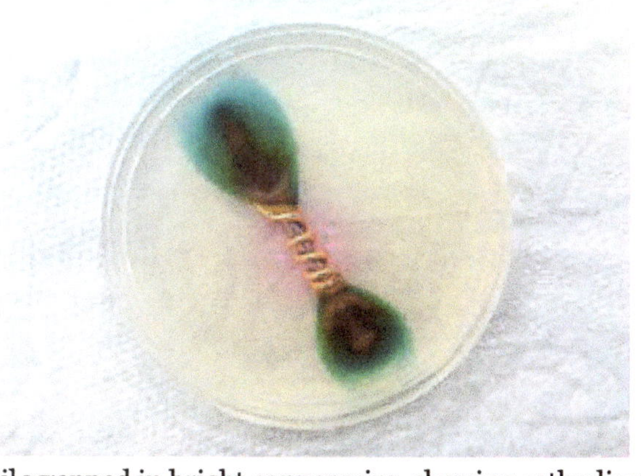

Corrosion of an iron nail wrapped in bright copper wire, showing cathodic protection of copper; a ferroxyl indicator solution shows colored chemical indications of two types of ions diffusing through a moist agar medium

Overview

Dissimilar metals and alloys have different electrode potentials, and when two or more come into contact in an electrolyte, one metal acts as anode and the other as cathode. The electropotential difference between the dissimilar metals is the driving force for an accelerated attack on the anode member of the galvanic couple. The anode metal dissolves into the electrolyte, and deposit collects on the cathodic metal.

The electrolyte provides a means for ion migration whereby metallic ions move from the anode to the cathode within the metal. This leads to the metal at the anode corroding more quickly than it otherwise would and corrosion at the cathode being inhibited. The presence of an electrolyte and an electrical conducting path between the metals is essential for galvanic corrosion to occur.

In some cases, this type of reaction is intentionally encouraged. For example, low-cost household batteries typically contain carbon-zinc cells. As part of a closed circuit (the electron pathway), the zinc within the cell will corrode preferentially (the ion pathway) as an essential part of the battery producing electricity. Another example is the cathodic protection of buried or submerged structures. In this case, sacrificial anodes work as part of a galvanic couple, promoting corrosion of the anode, while protecting the cathode metal.

In other cases, such as mixed metals in piping (for example, copper, cast iron and other cast metals), galvanic corrosion will contribute to accelerated corrosion of parts of the system. Corrosion inhibitors such as sodium nitrite or sodium molybdate can be injected into these systems to reduce the galvanic potential. However, the application of these corrosion inhibitors must be monitored closely. If the application of corrosion inhibitors increases the conductivity of the water within the system, the galvanic corrosion potential can be greatly increased.

Acidity or alkalinity (pH) is also a major consideration with regard to closed loop bimetallic circulating systems. Should the pH and corrosion inhibition doses be incorrect, galvanic corrosion will be accelerated. In most HVAC systems, the use of sacrificial anodes and cathodes is not an option, as they would need to be applied within the plumbing of the system and, over time, would corrode and release particles that could cause potential mechanical damage to circulating pumps, heat exchangers, etc.

Examples of Corrosion

A common example of galvanic corrosion is the rusting of corrugated iron sheet, which becomes widespread when the protective zinc coating is broken and the underlying steel is attacked. The zinc is attacked preferentially because it is less noble, but once it has been consumed, rusting of the base metal can occur in earnest. By contrast, with a traditional tin can, the opposite of a protective effect occurs: because the tin is more noble than the underlying steel, when the tin coating is broken, the steel beneath is immediately attacked preferentially.

Statue of Liberty

A spectacular example of galvanic corrosion occurred in the Statue of Liberty when regular maintenance checks in the 1980s revealed that corrosion had taken place between the outer copper skin and the wrought iron support structure. Although the problem had been anticipated when the structure was built by Gustave Eiffel to Frédéric Bartholdi's design in the 1880s, the insulation layer of shellac between the two metals had failed over time and resulted in rusting of the iron supports. An extensive renovation requiring complete disassembly of the statue replaced the original insulation with PTFE. The structure was far from unsafe owing to the large number of unaffected connections, but it was regarded as a precautionary measure to preserve a national symbol of the United States.

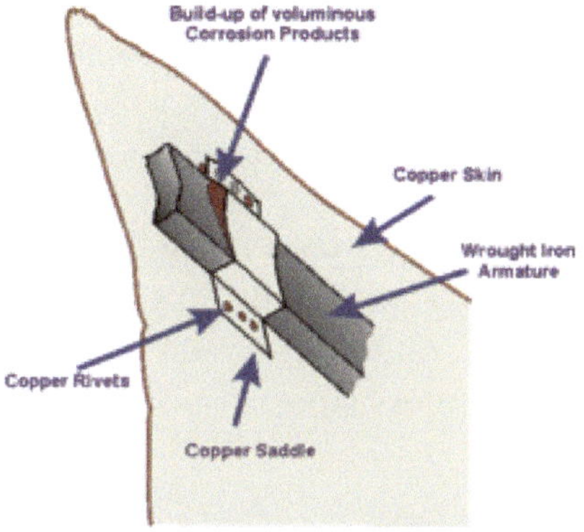

Galvanic corrosion in the Statue of Liberty

Regular maintenance checks discovered that the Statue of Liberty suffered from galvanic corrosion

Royal Navy and HMS Alarm

In 17th-century England, Samuel Pepys (then serving as Admiralty Secretary), agreed to the removal of lead sheathing from British Royal Navy vessels to prevent the mysterious disintegration of their rudder-irons and bolt-heads, though he confessed himself baffled as to the reason the lead caused the corrosion.

The problem recurred when vessels were sheathed in copper to reduce marine weed accumulation and protect against shipworm. In an experiment, the Royal Navy in 1761 had tried fitting the hull of the frigate HMS *Alarm* with 12-ounce copper plating. Upon her return from a voyage to the West Indies, it was found that although the copper remained in fine condition and had indeed deterred shipworm, it had also become detached from the wooden hull in many places because the iron nails used during its installation "...were found dissolved into a kind of rusty Paste". To the surprise of the inspection teams, however, some of the iron nails were virtually undamaged. Closer inspection revealed that water-resistant brown paper trapped under the nail head had inadvertently protected some of the nails: "Where this covering was perfect, the Iron was preserved from Injury". The copper sheathing had been delivered to the dockyard wrapped in the paper which was not always removed before the sheets were nailed to the hull. The conclusion therefore reported to the Admiralty in 1763 was that iron should not be allowed direct contact with copper in sea water.

US Navy Littoral Combat Ship Independence

Serious galvanic corrosion has been reported on the latest US Navy attack littoral combat vessel the USS *Independence* caused by steel water jet propulsion systems attached to an aluminium hull. Without electrical isolation between the steel and aluminium, the aluminium hull acts as an anode to the stainless steel, resulting in aggressive galvanic corrosion.

Lasagna Cell

A "lasagna cell" is accidentally produced when salty moist food such as lasagna is stored in a steel baking pan and is covered with aluminum foil. After a few hours the foil develops small holes where it touches the lasagna, and the food surface becomes covered with small spots composed of corroded aluminum.

In this example, the salty food (lasagna) is the electrolyte, the aluminum foil is the anode, and the steel pan is the cathode. If the aluminum foil only touches the electrolyte in small areas, the galvanic corrosion is concentrated, and corrosion can occur fairly rapidly.

Electrolytic Cleaning

The common technique of cleaning silverware by immersion of the silver and a piece of aluminum in an electrolytic bath (usually sodium bicarbonate) is an example of galvanic corrosion. (Care should be exercised because this will also strip silver oxide from the silverware, which may be there for decoration. Use on silverplate is inadvisable, as this may cause unwanted galvanic corrosion of the base metal.)

Preventing Galvanic Corrosion

Aluminum anodes mounted on a steel-jacketed structure

Electrical panel for a cathodic protection system

There are several ways of reducing and preventing this form of corrosion.

- Electrically insulate the two metals from each other. If they are not in electrical contact, no galvanic coupling will occur. This can be achieved by using non-conductive materials between metals of different electropotential. Piping can be isolated with a spool of pipe made of plastic materials, or made of metal material internally coated or lined. It is important that the spool be a sufficient length to be effective. For reasons of safety, this should not be attempted where an electrical earthing system uses the pipework for its ground or has equipotential bonding.
- Metal boats connected to a shore line electrical power feed will normally have to have the hull connected to earth for safety reasons. However the end of that earth connection is likely to be a copper rod buried within the marina, resulting in a steel-copper "battery" of about 0.5V. For such cases, the use of a galvanic isolator is essential, typically two semiconductor diodes in series, in parallel with two diodes conducting in the opposite direction. This prevents any current flow while the applied voltage is less than 1.4V (i.e. 0.7V per diode), but allows a full flow of current in case of an electrical fault. There will still be a very minor leakage of current through the diodes, which may result in slightly faster corrosion than normal.
- Ensure there is no contact with an electrolyte. This can be done by using water-repellent compounds such as greases, or by coating the metals with an impermeable protective layer, such as a suitable paint, varnish, or plastic. If it is not possible to coat both, the coating should be applied to the more noble, the material with higher potential. This is advisable because if the coating is applied only on the more active material, in case of damage to the coating there will be a large cathode area and a very small anode area, and for the exposed anodic area the corrosion rate will be correspondingly high.
- Using antioxidant paste is beneficial for preventing corrosion between copper and aluminum electrical connections. The paste consists of a lower nobility metal than aluminum or copper.

- Choose metals that have similar electropotentials. The more closely matched the individual potentials, the lesser the potential difference and hence the lesser the galvanic current. Using the same metal for all construction is the easiest way of matching potentials.
- Electroplating or other plating can also help. This tends to use more noble metals that resist corrosion better. Chrome, nickel, silver and gold can all be used. Galvanizing with zinc protects the steel base metal by sacrificial anodic action.
- Cathodic protection uses one or more sacrificial anodes made of a metal which is more active than the protected metal. Alloys of metals commonly used for sacrificial anodes include zinc, magnesium, and aluminium. This approach is commonplace in water heaters and many buried or immersed metallic structures.
- Cathodic Protection can also be applied by connecting a direct current (DC) electrical power supply to oppose the corrosive galvanic current.

Galvanic Series

All metals can be classified into a galvanic series representing the electrical potential they develop in a given electrolyte against a standard reference electrode. The relative position of two metals on such a series gives a good indication of which metal is more likely to corrode more quickly. However, other factors such as water aeration and flow rate can influence the rate of the process markedly.

Stainless steel cable ladder with mild steel bolts

Anodic Index

The compatibility of two different metals may be predicted by consideration of their anodic index. This parameter is a measure of the electrochemical voltage that will be developed between the metal and gold. To find the relative voltage of a pair of metals it is only required to subtract their anodic indices.

For normal environments, such as storage in warehouses or non-temperature and humidity controlled environments, there should not be more than 0.25 V difference in the anodic index. For controlled environments, in which temperature and humidity are controlled, 0.50 V can be tolerated. For harsh environments, such as outdoors, high humidity, and salt environments, there should be not more than 0.15 V difference in the anodic index. For example: gold/silver would have a difference of 0.15V, being acceptable in a harsh environment.

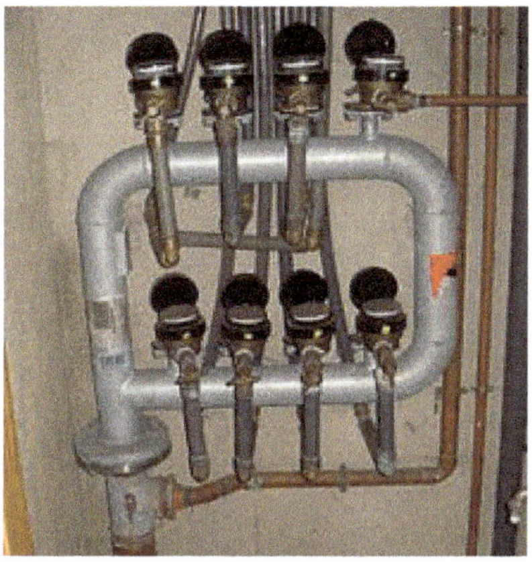

This new manifold for water meters has replaced the old one after 4 years of use, having been connected directly to a copper pipe in the building (0.85 V of potential difference, see the point of corrosion in the bottom left). With a PVC battery support, lifetime is unlimited.

Zinc sacrificial anode (photo center) to protect a boat

Often when design requires that dissimilar metals come in contact, the galvanic compatibility is managed by finishes and plating. The finishing and plating selected allows the dissimilar materials to be in contact, while protecting the base materials from corrosion. Note that it will always be the metal with the most negative anodic index which will ultimately suffer from corrosion when galvanic incompatibility is in play. This is why you should never place sterling silver and stainless steel tableware in a dishwasher at the same time, as the steel items will likely experience corrosion by the end of the cycle (soap and water having served as the chemical electrolyte, and heat having amplified the process).

Anodic index	
Metal	**Index (V)**
Most Cathodic	
Gold, solid and plated, Gold-platinum alloy	−0.00

Rhodium plated on silver-plated copper	−0.05
Silver, solid or plated; monel metal. High nickel-copper alloys	−0.15
Nickel, solid or plated, titanium an s alloys, Monel	−0.30
Copper, solid or plated; low brasses or bronzes; silver solder; German silvery high copper-nickel alloys; nickel-chromium alloys	−0.35
Brass and bronzes	−0.40
High brasses and bronzes	−0.45
18% chromium type corrosion-resistant steels	−0.50
Chromium plated; tin plated; 12% chromium type corrosion-resistant steels	−0.60
Tin-plate; tin-lead solder	−0.65
Lead, solid or plated; high lead alloys	−0.70
2000 series wrought aluminum	−0.75
Iron, wrought, gray or malleable, plain carbon and low alloy steels	−0.85
Aluminum, wrought alloys other than 2000 series aluminum, cast alloys of the silicon type	−0.90
Aluminum, cast alloys other than silicon type, cadmium, plated and chromate	−0.95
Hot-dip-zinc plate; galvanized steel	−1.20
Zinc, wrought; zinc-base die-casting alloys; zinc plated	−1.25
Magnesium & magnesium-base alloys, cast or wrought	−1.75
Beryllium	−1.85
Most Anodic	

Galvanization

A street lamp in Singapore showing the characteristic spangle of hot-dip galvanization.

Galvanization or galvanisation (or galvanizing as it is most commonly called in that industry) is the process of applying a protective zinc coating to steel or iron, to prevent rusting. The most common method is hot-dip galvanizing, in which parts are submerged in a bath of molten zinc. Galvanizing protects in three ways:

- It forms a coating of zinc which, when intact, prevents corrosive substances from reaching the underlying steel or iron.
- The zinc serves as a sacrificial anode so that even if the coating is scratched, the exposed steel will still be protected by the remaining zinc.
- The zinc protects its base metal by corroding before iron. For better results, application of chromates over zinc is also seen as an industrial trend.

Etymology

The earliest known example of galvanizing of iron, encountered by Europeans is found on 17th-century Indian armor in the Royal Armouries Museum collection. It was named in English via French from the name of Italian scientist Luigi Galvani. Originally, galvanizing was the administration of electric shocks, in the 19th century also termed Faradism. This sense is the origin of the meaning of the metaphorical use of the verb 'galvanize', as in 'galvanize into action', or to stimulate a complacent person or group to take action. The term galvanizing has largely come to be associated with zinc coatings, to the exclusion of other metals. Galvanic paint, a precursor to hot-dip galvanizing, was patented by Stanislas Sorel, of Paris in December, 1837.

Galvanized nails.

Methods

Hot-dip galvanizing deposits a thick robust layer of zinc iron alloys on the surface of a steel item. In the case of automobile bodies, where additional decorative coatings of paint will be applied, a thinner form of galvanizing is applied by electrogalvanizing. The hot-dip process generally does not reduce strength on a measurable scale, with the exception of high-strength steels (>1100 MPa) where hydrogen embrittlement can become a problem. This is a consideration for the manufacture of wire rope and other highly stressed products. The protection provided by hot-dip galvanizing is insufficient for products that will be constantly exposed to corrosive materials such as acids. For these applications, more expensive stainless steel is preferred. Some nails made today are galvanized. Nonetheless, electroplating is used on its own for many outdoor applications because it is cheaper than hot-dip zinc coating and looks good when new.

Another reason not to use hot-dip zinc coating is that for bolts and nuts size M10 (US 3/8") or smaller, the thick hot-dipped coating fills in too much of the threads, which reduces strength (because the dimension of the steel prior to coating must be reduced for the fasteners to fit together). This means that for cars, bicycles, and many other light mechanical products, the alternative to electroplating bolts and nuts is not hot-dip zinc coating, but making the bolts and nuts from stainless steel.

Galvanized surface with visible spangle

The size of crystallites in galvanized coatings is a visible and aesthetic feature, known as "spangle". By varying the number of particles added for heterogeneous nucleation and the rate of cooling in a hot-dip process, the spangle can be adjusted from an apparently uniform surface to grains several centimetres wide. Visible crystallites are rare in other engineering materials.

Thermal diffusion galvanizing, or Sherardizing, provides a zinc diffusion coating on iron- or copper-based materials. Parts and zinc powder are tumbled in a sealed rotating drum. Around 300°C, zinc will diffuse into the substrate to form a zinc alloy. The preparation of the goods can be carried out by shot blasting. The process is also known as dry galvanizing, because no liquids are involved; however, no danger of hydrogen embrittlement of the goods exists. The dull-grey crystal structure of the zinc diffusion coating has a good adhesion to paint, powder coatings, or rubber. It is a preferred method for coating small, complex-shaped metals, and for smoothing rough surfaces on items formed with powder metal.

Eventual Corrosion

Although galvanizing will inhibit attack of the underlying steel, rusting will be inevitable, after some decades, especially if exposed to acidic conditions. For example, corrugated iron sheet roofing will start to degrade within a few years despite the protective action of the zinc coating. Marine and salty environments also lower the lifetime of galvanized iron because the high electrical conductivity of sea water increases the rate of corrosion primarily through converting the solid zinc to soluble zinc chloride which simply washes away. Galvanized car frames exemplify this; they corrode much quicker in cold environments due to road salt, though they will last longer than unprotected steel. Galvanized steel can last for many decades if other means are maintained, such as paint coatings and additional sacrificial anodes. The rate of corrosion in non-salty environments is mainly due to levels of sulfur dioxide in the air. In the most benign natural environments, such as inland low population areas, galvanized steel can last without rust for over 100 years.

Rusted corrugated steel roof

Galvanized Piping

In the early 20th century, galvanized piping replaced cast iron and lead in cold-water plumbing. Typically, galvanized piping rusts from the inside out, building up plaques on the inside of the piping, causing both water pressure problems and eventual pipe failure. These plaques can flake off, leading to visible impurities in water and a slight metallic taste. The life expectancy of such piping is about 70 years, but it may vary by region due to impurities in the water supply and the proximity of electrical grids for which interior piping acts as a pathway (the flow of electricity can accelerate chemical corrosion). Pipe longevity also depends on the thickness of zinc in the original galvanizing, which ranges on a scale from G40 to G210, and whether the pipe was galvanized on both the inside and outside, or just the outside. Since World War II, copper and plastic piping have replaced galvanized piping for interior drinking water service, but galvanized steel pipes are still used in outdoor applications requiring steel's superior mechanical strength.

This lends some truth to the urban myth that water purity in outdoor water faucets is lower, but the actual impurities (iron, zinc, calcium) are harmless.

The presence of galvanized piping detracts from the appraised value of housing stock because piping can fail, increasing the risk of water damage. Galvanized piping will eventually need to be replaced if housing stock is to outlast a 50 to 70 year life expectancy, and some jurisdictions require galvanized piping to be replaced before sale. One option to extend the life expectancy of existing galvanized piping is to line it with an epoxy resin.

Galvanized Construction Steel

This is the most common use for galvanizing, and hundreds of thousands of tonnes are galvanized annually worldwide. In developed countries most larger cities have several galvanizing factories, and many items of steel manufacture are galvanized for protection. Typically these include: street furniture, building frameworks, balconies, verandahs, staircases, ladders, walkways and more.

Electrogalvanization

Electrogalvaning is a process in which a layer of zinc is bonded to steel in order to protect against corrosion. The process involves electroplating, running a current of electricity through a saline/zinc solution with a zinc anode and steel conductor. Zinc electroplating maintains a dominant position among other electroplating process options, based upon electroplated tonnage per annum. According to the International Zinc Association, more than 5 million tons are used yearly for both Hot Dip Galvanizing and Electroplating. The Plating of Zinc was developed at the beginning of the 20th century. At that time, the electrolyte was cyanide based. A significant innovation occurred in the 1960s, with the introduction of the first acid chloride based electrolyte. The 1980s saw a return to alkaline electrolytes, only this time, without the use of cyanide. Compared to hot dip galvanizing, electroplated zinc offers these significant advantages:

- Lower thickness deposits to achieve comparable performance
- Broader conversion coating availability for increased performance and colour options
- Brighter, more aesthetically appealing, deposits

History

Zinc plating was developed and continues to evolve, to meet the most challenging corrosion protection, temperature and wear resistance requirements. Electroplating of zinc was invented in 1800 but the first bright deposits were not obtained until the early 1930s with the alkaline cyanide electrolyte. Much later, in 1966, the use of acid chloride baths improved the brightness even greater. The latest modern development occurred in the 1980s, with the new generation of alkaline, cyanide-free zinc. Recent European Union directives (ELV/RoHS/WEEE) prohibit automotive, other original equipment manufacturers (OEM) and electrical and electronic equipment manufacturers from using hexavalent chromium (CrVI). These directives combined with increased performance requirements by the OEM, has led to an increase in the use of alkaline zinc, zinc alloys and high performance trivalent passivate conversion coatings.

Processes

The corrosion protection afforded by the electrodeposited zinc layer is primarily due to the anodic potential dissolution of zinc versus iron (the substrate in most cases). Zinc acts as a sacrificial anode for protecting the iron (steel). While steel is close to E_{SCE}= -400 mV (the potential refers to the standard Saturated calomel electrode (SCE), depending on the alloy composition, electroplated zinc is much more anodic with E_{SCE}= -980 mV. Steel is preserved from corrosion by cathodic protection. Conversion coatings (hexavalent chromium (CrVI) or trivalent chromium (CrIII) depending upon OEM requirements) are applied to drastically enhance the corrosion protection by building an additional inhibiting layer of Chromium and Zinc hydroxides. These oxide films range in thickness from 10 nm for the thinnest blue/clear passivates to 4 µm for the thickest black chromates.

Additionally, electroplated zinc articles may receive a topcoat to further enhance corrosion protection and friction performance.

The modern electrolytes are both alkaline and acidic:

Alkaline Electrolytes

Cyanide Electrolytes

Contain sodium sulphate and sodium hydroxide (NaOH). All of them utilize proprietary brightening agents. Zinc is soluble as a cyanide complex $Na_2Zn(CN)_4$ and as a zincate $Na_2Zn(OH)_4$. Quality control of such electrolytes requires the regular analysis of Zn, NaOH and NaCN. The ratio of NaCN : Zn can vary between 2 and 3 depending upon the bath temperature and desired deposit brightness level.The following chart illustrates the typical cyanide electrolyte options used to plate at room temperature:

Cyanide bath composition			
	Zinc	**Sodium hydroxide**	**Sodium cyanide**
Low cyanide	6-10 g/L (0.8-1.3 oz/gal)	75-90 g/L (10-12 oz/gal)	10-20 g/L 1.3-2.7 oz/gal)
Mid cyanide	15-20 g/L (2.0-2.7 oz/gal)	75-90 g/L (10-12 oz/gal)	25-45 g/L (3.4-6.0 oz/gal)
High cyanide	25-35 g/L (3.4-4.7 oz/gal)	75-90 g/L (10-12 oz/gal)	80-100 g/L (10.70- 13.4 oz/gal)

Alkaline Non-cyanide Electrolytes

Contain zinc and sodium hydroxide. Most of them are brightened by proprietary addition agents similar to those used in cyanide baths. The addition of quaternary amine additives contribute to the improved metal distribution between high and low current density areas. Depending upon the desired performance, the electroplater can select the highest zinc content for increased productivity or lower zinc content for a better throwing power (into low current density areas). For ideal metal distribution, Zn metal evolutes between 6-14 g/L (0.8-1.9 oz/gal) and NaOH at 120 g/L (16 oz/gal). But for the highest productivity, Zn metal is between 14-25 g/L (1.9-3.4 oz/gal) and

NaOH remains at 120 g/L (16 oz/gal). Alkaline Non Cyanide Zinc Process contains lower concentration zinc metal concentration 6-14 g/L (0.8-1.9 oz/gal) or higher zinc metal concentration 14-25 g/L (1.9-3.4 oz/gal) provides superior plate distribution from high current density to low current density or throwing power when compared to any acidic baths such as chloride based (Low ammonium chloride, Potassium chloride / Ammonium Chloride) - or (non-ammonium chloride, potassium chloride/Boric acid) or sulfate baths.

Acidic Electrolytes

High Speed Electrolytes

Dedicated to plating at high speed in plants where the shortest plating time is critical (i.e. steel coil or pipe that runs at up to 200 m/min. The baths contain zinc sulfate and chloride to the maximum solubility level. Boric acid may be used as a pH buffer and to reduce the burning effect at high current densities. These baths contain very few grain refiners. If one is utilized, it may be sodium saccharine.

Traditional Electrolytes

Initially based on ammonium chloride, options today include ammonium, potassium or mixed ammonium/potassium electrolytes. The chosen content of zinc depends on the required productivity and part configuration. High zinc improves the bath's efficiency (plating speed), while lower levels improve the bath's ability to throw into low current densities. Typically, the Zn metal level varies between 20 and 50 g/L (2.7-6.7 oz/gal). The pH varies between 4.8 and 5.8 units. The following chart illustrates a typical all potassium chloride bath composition:

Traditional acid bath composition	
Parameters	**Value in g/L (oz/gal)**
Zinc	40 g/l (5.4 oz/gal)
Total chloride	125 g/l (16.8 oz/gal)
Anhydrous zinc chloride	80 g/l (10.7 oz/gal)
Potassium chloride	180 g/l (24.1 oz/gal)
Boric acid	25 g/l (3.4 oz/gal)

Typical grain refiners include low soluble ketones and aldehydes. These brightening agents must be dissolved in alcohol or in hydrotrope. The resultant molecules are co-deposited with the zinc to produce a slightly leveled, very bright deposit. The bright deposit has also been shown to decrease chromate/passivate receptivity, however. The result is a reduction in the corrosion protection afforded.

Electroplating

Electroplating is a process that uses electric current to reduce dissolved metal cations so that they form a coherent metal coating on an electrode. The term is also used for electrical oxidation of anions onto a solid substrate, as in the formation silver chloride on silver wire to make silver/silver-chloride electrodes. Electroplating is primarily used to change the surface properties of an object (e.g. abrasion and wear resistance, corrosion protection, lubricity, aesthetic qualities, etc.), but may also be used to build up thickness on undersized parts or to form objects by electroforming.

Copper electroplating machine for layering PCBs

The process used in electroplating is called electrodeposition. It is analogous to a galvanic cell acting in reverse. The part to be plated is the cathode of the circuit. In one technique, the anode is made of the metal to be plated on the part. Both components are immersed in a solution called an electrolyte containing one or more dissolved metal salts as well as other ions that permit the flow of electricity. A power supply supplies a direct current to the anode, oxidizing the metal atoms that it comprises and allowing them to dissolve in the solution. At the cathode, the dissolved metal ions in the electrolyte solution are reduced at the interface between the solution and the cathode, such that they "plate out" onto the cathode. The rate at which the anode is dissolved is equal to the rate at which the cathode is plated, vis-a-vis the current through the circuit. In this manner, the ions in the electrolyte bath are continuously replenished by the anode.

Other electroplating processes may use a non-consumable anode such as lead or carbon. In these techniques, ions of the metal to be plated must be periodically replenished in the bath as they are drawn out of the solution. The most common form of electroplating is used for creating coins such as pennies, which are small zinc plates covered in a layer of copper.

Process

The cations associate with the anions in the solution. These cations are reduced at the cathode to deposit in the metallic, zero valence state. For example, for copper plating, in an acid solution, copper is oxidized at the anode to Cu^{2+} by losing two electrons. The Cu^{2+} associates with the anion SO_4^{2-} in the solution to form copper sulfate. At the cathode, the Cu^{2+} is reduced to metallic copper by gaining two electrons. The result is the effective transfer of copper from the anode source to a plate covering the cathode.

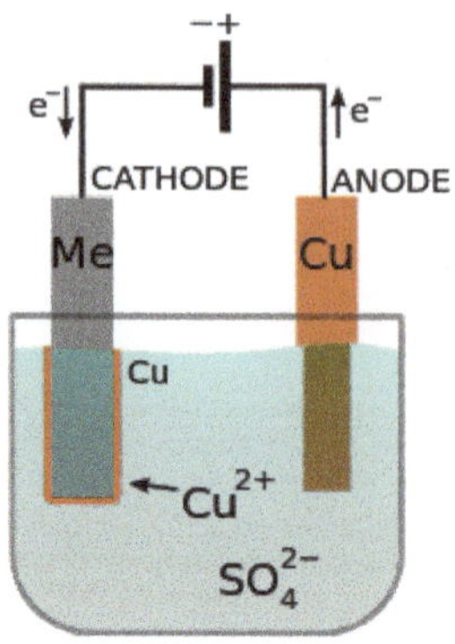

Electroplating of a metal (Me) with copper in a copper sulfate bath

The plating is most commonly a single metallic element, not an alloy. However, some alloys can be electrodeposited, notably brass and solder.

Many plating baths include cyanides of other metals (e.g., potassium cyanide) in addition to cyanides of the metal to be deposited. These free cyanides facilitate anode corrosion, help to maintain a constant metal ion level and contribute to conductivity. Additionally, non-metal chemicals such as carbonates and phosphates may be added to increase conductivity.

When plating is not desired on certain areas of the substrate, stop-offs are applied to prevent the bath from coming in contact with the substrate. Typical stop-offs include tape, foil, lacquers, and waxes.

The ability of a plating to cover uniformly is called *throwing power*; the better the "throwing power" the more uniform the coating.

Strike

Initially, a special plating deposit called a "strike" or "flash" may be used to form a very thin (typically less than 0.1 micrometer thick) plating with high quality and good adherence to the substrate. This serves as a foundation for subsequent plating processes. A strike uses a high current density and a bath with a low ion concentration. The process is slow, so more efficient plating processes are used once the desired strike thickness is obtained.

The striking method is also used in combination with the plating of different metals. If it is desirable to plate one type of deposit onto a metal to improve corrosion resistance but this metal has inherently poor adhesion to the substrate, a strike can be first deposited that is compatible with both. One example of this situation is the poor adhesion of electrolytic nickel on zinc alloys, in which case a copper strike is used, which has good adherence to both.

Electrochemical Deposition

Electrochemical deposition is generally used for the growth of metals and conducting metal oxides because of the following advantages: (i) the thickness and morphology of the nanostructure can be precisely controlled by adjusting the electrochemical parameters, (ii) relatively uniform and compact deposits can be synthesized in template-based structures, (iii) higher deposition rates are obtained, and (iv) the equipment is inexpensive due to the non-requirements of either a high vacuum or a high reaction temperature.

Pulse Electroplating or Pulse Electrodeposition (PED)

A simple modification in the electroplating process is the pulse electroplating. This process involves the swift alternating of the potential or current between two different values resulting in a series of pulses of equal amplitude, duration and polarity, separated by zero current. By changing the pulse amplitude and width, it is possible to change the deposited film's composition and thickness.

Brush Electroplating

A closely related process is brush electroplating, in which localized areas or entire items are plated using a brush saturated with plating solution. The brush, typically a stainless steel body wrapped with a cloth material that both holds the plating solution and prevents direct contact with the item being plated, is connected to the positive side of a low voltage direct-current power source, and the item to be plated connected to the negative. The operator dips the brush in plating solution then applies it to the item, moving the brush continually to get an even distribution of the plating material. Brush electroplating has several advantages over tank plating, including portability, ability to plate items that for some reason cannot be tank plated (one application was the plating of portions of very large decorative support columns in a building restoration), low or no masking requirements, and comparatively low plating solution volume requirements. Disadvantages compared to tank plating can include greater operator involvement (tank plating can frequently be done with minimal attention), and inability to achieve as great a plate thickness.

Electroless Deposition

Usually an electrolytic cell (consisting of two electrodes, electrolyte, and external source of current) is used for electrodeposition. In contrast, an electroless deposition process uses only one electrode and no external source of electric current. However, the solution for the electroless process needs to contain a reducing agent so that the electrode reaction has the form:

$$M^{z+} + Red_{solution} \overset{\text{catalytic surface}}{\Rightarrow} M_{solid} + Oxy_{solution}$$

In principle any hydrogen-based reducer can be used although the redox potential of the reducer half-cell must be high enough to overcome the energy barriers inherent in liquid chemistry. Electroless nickel plating uses hypophosphite as the reducer while plating of other metals like silver, gold and copper typically use low molecular weight aldehydes.

A major benefit of this approach over electroplating is that the power sources and plating baths are not needed, reducing the cost of production. The technique can also plate diverse shapes and types of surface. The downside is that the plating process is usually slower and cannot create such thick plates of metal. As a consequence of these characteristics, electroless deposition is quite common in the decorative arts.

Cleanliness

Cleanliness is essential to successful electroplating, since molecular layers of oil can prevent adhesion of the coating. ASTM B322 is a standard guide for cleaning metals prior to electroplating. Cleaning processes include solvent cleaning, hot alkaline detergent cleaning, electro-cleaning, and acid treatment etc. The most common industrial test for cleanliness is the waterbreak test, in which the surface is thoroughly rinsed and held vertical. Hydrophobic contaminants such as oils cause the water to bead and break up, allowing the water to drain rapidly. Perfectly clean metal surfaces are hydrophilic and will retain an unbroken sheet of water that does not bead up or drain off. ASTM F22 describes a version of this test. This test does not detect hydrophilic contaminants, but the electroplating process can displace these easily since the solutions are water-based. Surfactants such as soap reduce the sensitivity of the test and must be thoroughly rinsed off.

Effects

Electroplating changes the chemical, physical, and mechanical properties of the workpiece. An example of a chemical change is when nickel plating improves corrosion resistance. An example of a physical change is a change in the outward appearance. An example of a mechanical change is a change in tensile strength or surface hardness which is a required attribute in tooling industry. Electroplating of acid gold on underlying copper/nickel-plated circuits reduces contact resistance as well as surface hardness. Copper-plated areas of mild steel act as a mask if case hardening of such areas are not desired. Tin-plated steel is chromium-plated to prevent dulling of the surface due to oxidation of tin.

History

Modern electrochemistry was invented by Italian chemist Luigi Valentino Brugnatelli (it) in 1805. Brugnatelli used his colleague Alessandro Volta's invention of five years earlier, the voltaic pile, to facilitate the first electrodeposition. Brugnatelli's inventions were suppressed by the French Academy of Sciences and did not become used in general industry for the following thirty years.

Luigu Valentino Brugnatelli

Nickel plating

By 1839, scientists in Britain and Russia had independently devised metal deposition processes similar to Brugnatelli's for the copper electroplating of printing press plates.

Boris Jacobi developed electroplating, electrotyping and galvanoplastic sculpture in Russia

Boris Jacobi in Russia not only rediscovered galvanoplastics, but developed electrotyping and galvanoplastic sculpture. Galvanoplastics quickly came into fashion in Russia, with such people as inventor Peter Bagration, scientist Heinrich Lenz and science fiction author Vladimir Odoyevsky all contributing to further development of the technology. Among the most notorious cases of electroplating usage in mid-19th century Russia were gigantic galvanoplastic sculptures of St. Isaac's Cathedral in Saint Petersburg and gold-electroplated dome of the Cathedral of Christ the Saviour in Moscow, the tallest Orthodox church in the world.

Galvanoplastic sculpture on St. Isaac's Cathedral in Saint Petersburg.

Soon after, John Wright of Birmingham, England discovered that potassium cyanide was a suitable electrolyte for gold and silver electroplating. Wright's associates, George Elkington and Henry Elkington were awarded the first patents for electroplating in 1840. These two then founded the electroplating industry in Birmingham from where it spread around the world. The Woolrich Electrical Generator of 1844, now in Thinktank, Birmingham Science Museum, is the earliest electrical generator used in an industrial process. It was used by Elkingtons.

The Woolrich Electrical Generator in Thinktank, Birmingham

The Norddeutsche Affinerie in Hamburg was the first modern electroplating plant starting its production in 1876.

As the science of electrochemistry grew, its relationship to the electroplating process became understood and other types of non-decorative metal electroplating processes were developed. Commercial electroplating of nickel, brass, tin, and zinc were developed by the 1850s. Electroplating baths and equipment based on the patents of the Elkingtons were scaled up to accommodate the plating of numerous large scale objects and for specific manufacturing and engineering applications.

The plating industry received a big boost with the advent of the development of electric generators in the late 19th century. With the higher currents available, metal machine components, hardware, and automotive parts requiring corrosion protection and enhanced wear properties, along with better appearance, could be processed in bulk.

The two World Wars and the growing aviation industry gave impetus to further developments and refinements including such processes as hard chromium plating, bronze alloy plating, sulfamate nickel plating, along with numerous other plating processes. Plating equipment evolved from manually operated tar-lined wooden tanks to automated equipment, capable of processing thousands of kilograms per hour of parts.

One of the American physicist Richard Feynman's first projects was to develop technology for electroplating metal onto plastic. Feynman developed the original idea of his friend into a successful invention, allowing his employer (and friend) to keep commercial promises he had made but could not have fulfilled otherwise.

Uses

Electroplating is widely used in various industries for coating metal objects with a thin layer of a different metal. The layer of metal deposited has some desired property, which the metal of the object lacks. For example, chromium plating is done on many objects such as car parts, bath taps, kitchen gas burners, wheel rims and many others for the fact that chromium is very corrosion resistant, and thus prolongs the life of the parts. Electroplating has wide usage in industries. It is also used in making inexpensive jewelry. Electroplating increases life of metal and prevents corrosion.

A zinc solution tested in a hull cell

Hull Cell

The *Hull cell* is a type of test cell used to qualitatively check the condition of an electroplating bath. It allows for optimization for current density range, optimization of additive concentration, recognition of impurity effects and indication of macro-throwing power capability. The Hull cell replicates the plating bath on a lab scale. It is filled with a sample of the plating solution, an appropriate anode which is connected to a rectifier. The "work" is replaced with a hull cell test panel that will be plated to show the "health" of the bath.

The Hull cell is a trapezoidal container that holds 267 ml of solution. This shape allows one to place the test panel on an angle to the anode. As a result, the deposit is plated at different current densities which can be measured with a hull cell ruler. The solution volume allows for a quantitative optimization of additive concentration: 1 gram addition to 267 mL is equivalent to 0.5 oz/gal in the plating tank.

Haring-Blum Cell

The Haring-Blum Cell is used to determine the macro throwing power of a plating bath. The cell consists of 2 parallel cathodes with a fixed anode in the middle. The cathodes are at distances from the anode in the ratio of 1:5. The macro throwing power is calculating from the thickness of plating at the two cathodes when a dc current is passed for a specific period of time. The cell is fabricated out of Perspex or glass.

Redox

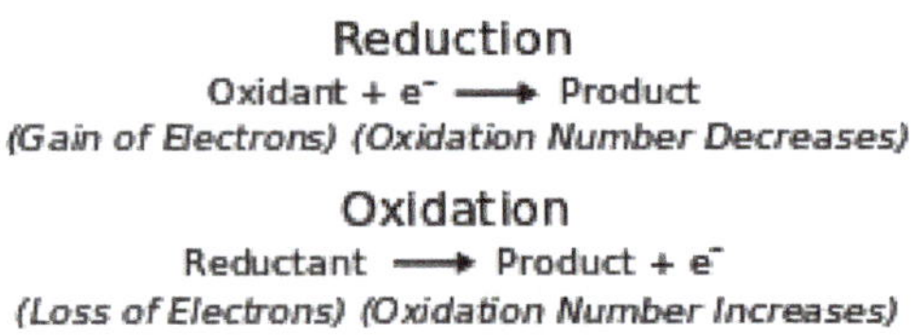

The two parts of a redox reaction

Rusting iron

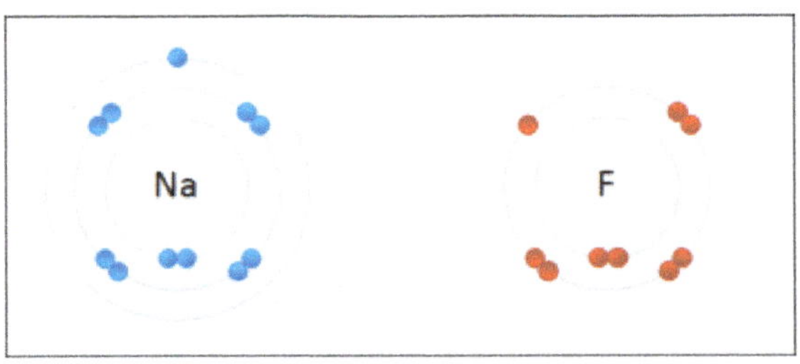

Sodium and fluorine bonding ionically to form sodium fluoride. Sodium loses its outer electron to give it a stable electron configuration, and this electron enters the fluorine atom exothermically. The oppositely charged ions are then attracted to each other. The sodium is oxidized, and the fluorine is reduced.

Redox (short for reduction–oxidation reaction) is a chemical reaction in which the oxidation states of atoms are changed. Any such reaction involves both a reduction process and a complementary oxidation process, two key concepts involved with electron transfer processes. Redox reactions include all chemical reactions in which atoms have their oxidation state changed; in general, redox reactions involve the transfer of electrons between chemical species. The chemical species from which the electron is stripped is said to have been oxidized, while the chemical species to which the electron is added is said to have been reduced. It can be explained in simple terms:

- Oxidation is the *loss* of electrons or an *increase* in oxidation state by a molecule, atom, or ion.
- Reduction is the *gain* of electrons or a *decrease* in oxidation state by a molecule, atom, or ion.

As an example, during the combustion of wood, oxygen from the air is reduced, transferring electrons from the carbon. Although oxidation reactions are commonly associated with the formation of oxides from oxygen molecules, oxygen is not necessarily included in such reactions, as other chemical species can serve the same function.

The reaction can occur relatively slowly, as in the case of rust, or more quickly, as in the case of fire. There are simple redox processes, such as the oxidation of carbon to yield carbon dioxide (CO_2) or the reduction of carbon by hydrogen to yield methane (CH_4), and more complex processes such as the oxidation of glucose ($C_6H_{12}O_6$) in the human body.

Etymology

"Redox" is a combination of "reduction" and "oxidation".

The word *oxidation* originally implied reaction with oxygen to form an oxide, since dioxygen (O_2 (g)) was historically the first recognized oxidizing agent. Later, the term was expanded to encompass oxygen-like substances that accomplished parallel chemical reactions. Ultimately, the meaning was generalized to include all processes involving loss of electrons.

The word *reduction* originally referred to the loss in weight upon heating a metallic ore such as a metal oxide to extract the metal. In other words, ore was "reduced" to metal. Antoine Lavoisier (1743–1794) showed that this loss of weight was due to the loss of oxygen as a gas. Later, scientists realized that the metal atom gains electrons in this process. The meaning of *reduction* then became generalized to include all processes involving gain of electrons. Even though "reduction" seems counter-intuitive when speaking of the gain of electrons, it might help to think of reduction as the loss of oxygen, which was its historical meaning. Since electrons are negatively charged, it is also helpful to think of this as reduction in electrical charge.

The electrochemist John Bockris has used the words *electronation* and *deelectronation* to describe reduction and oxidation processes respectively when they occur at electrodes. These words are analogous to protonation and deprotonation, but they have not been widely adopted by chemists.

The term "hydrogenation" could be used instead of reduction, since hydrogen is the reducing agent in a large number of reactions, especially in organic chemistry and biochemistry. But, unlike oxidation, which has been generalized beyond its root element, hydrogenation has maintained its specific connection to reactions that *add* hydrogen to another substance (e.g., the hydrogenation of unsaturated fats into saturated fats, R–CH=CH–R + H_2 → R–CH_2–CH_2–R). The word "redox" was first used in 1928.

Definitions

The processes of oxidation and reduction occur simultaneously and cannot happen independently of one another, similar to the acid–base reaction. The oxidation alone and the reduction alone are each called a *half-reaction*, because two half-reactions always occur together to form a whole reaction. When writing half-reactions, the gained or lost electrons are typically included explicitly in order that the half-reaction be balanced with respect to electric charge.

Though sufficient for many purposes, these general descriptions are not precisely correct. Although oxidation and reduction properly refer to *a change in oxidation state* — the actual transfer of electrons may never occur. The oxidation state of an atom is the fictitious charge that an atom would have if all bonds between atoms of different elements were 100% ionic. Thus, oxidation is best defined as an *increase in oxidation state*, and reduction as a *decrease in oxidation state*. In practice, the transfer of electrons will always cause a change in oxidation state, but there are many reactions that are classed as "redox" even though no electron transfer occurs (such as those involving covalent bonds).

Oxidizing and Reducing Agents

In redox processes, the reductant transfers electrons to the oxidant. Thus, in the reaction, the reductant or *reducing agent* loses electrons and is oxidized, and the oxidant or *oxidizing agent* gains electrons and is reduced. The pair of an oxidizing and reducing agent that are involved in a particular reaction is called a redox pair. A redox couple is a reducing species and its corresponding oxidizing form, e.g., Fe^{2+}/Fe^{3+}.

Oxidizers

Substances that have the ability to oxidize other substances (cause them to lose electrons) are said to be oxidative or oxidizing and are known as oxidizing agents, oxidants, or oxidizers. That is, the oxidant (oxidizing agent) removes electrons from another substance, and is thus itself reduced. And, because it "accepts" electrons, the oxidizing agent is also called an electron acceptor. Oxygen is the quintessential oxidizer.

The international pictogram for oxidising chemicals.

Oxidants are usually chemical substances with elements in high oxidation states (e.g., H2O2, MnO−4, CrO3, Cr2O2−7, OsO4), or else highly electronegative elements (O_2, F_2, Cl_2, Br_2) that can gain extra electrons by oxidizing another substance.

Reducers

Substances that have the ability to reduce other substances (cause them to gain electrons) are said to be reductive or reducing and are known as reducing agents, reductants, or reducers. The reductant (reducing agent) transfers electrons to another substance, and is thus itself oxidized. And, because it "donates" electrons, the reducing agent is also called an electron donor. Electron donors can also form charge transfer complexes with electron acceptors.

Reductants in chemistry are very diverse. Electropositive elemental metals, such as lithium, sodium, magnesium, iron, zinc, and aluminium, are good reducing agents. These metals donate or *give away* electrons readily. *Hydride transfer reagents*, such as $NaBH_4$ and $LiAlH_4$, are widely used in organic chemistry, primarily in the reduction of carbonyl compounds to alcohols. Another method of reduction involves the use of hydrogen gas (H_2) with a palladium, platinum, or nickel catalyst. These *catalytic reductions* are used primarily in the reduction of carbon-carbon double or triple bonds.

Standard Electrode Potentials (Reduction Potentials)

Each half-reaction has a *standard electrode potential* (*E*o cell), which is equal to the potential difference or voltage at equilibrium under standard conditions of an electrochemical cell in which the cathode reaction is the half-reaction considered, and the anode is a standard hydrogen electrode where hydrogen is oxidized:

$$\tfrac{1}{2} H_2 \rightarrow H^+ + e^-.$$

The electrode potential of each half-reaction is also known as its *reduction potential E*o red, or potential when the half-reaction takes place at a cathode. The reduction potential is a measure of the tendency of the oxidizing agent to be reduced. Its value is zero for $H^+ + e^- \rightarrow \tfrac{1}{2} H_2$ by definition, positive for oxidizing agents stronger than H^+ (e.g., +2.866 V for F_2) and negative for oxidizing agents that are weaker than H^+ (e.g., −0.763 V for Zn^{2+}).

For a redox reaction that takes place in a cell, the potential difference is:

> *E*o
cell = *E*o
cathode – *E*o
anode

However, the potential of the reaction at the anode was sometimes expressed as an *oxidation potential*:

> *E*o
ox = −*E*o
red.

The oxidation potential is a measure of the tendency of the reducing agent to be oxidized, but does not represent the physical potential at an electrode. With this notation, the cell voltage equation is written with a plus sign

> *E*o
cell = *E*o
red(cathode) + *E*o
ox(anode)

Examples of Redox Reactions

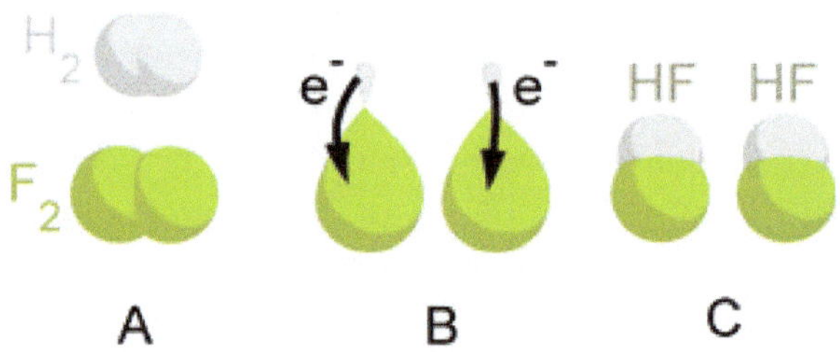

Illustration of a redox reaction

A good example is the reaction between hydrogen and fluorine in which hydrogen is being oxidized and fluorine is being reduced:

$$H2 + F2 \rightarrow 2HF$$

We can write this overall reaction as two half-reactions:

the oxidation reaction:

$$H2 \rightarrow 2H^+ + 2e^-$$

and the reduction reaction:

$$F2 + 2\,e^- \rightarrow 2\,F^-$$

Analyzing each half-reaction in isolation can often make the overall chemical process clearer. Because there is no net change in charge during a redox reaction, the number of electrons in excess in the oxidation reaction must equal the number consumed by the reduction reaction (as shown above).

Elements, even in molecular form, always have an oxidation state of zero. In the first half-reaction, hydrogen is oxidized from an oxidation state of zero to an oxidation state of +1. In the second half-reaction, fluorine is reduced from an oxidation state of zero to an oxidation state of −1.

When adding the reactions together the electrons are canceled:

$$H2 \rightarrow 2\ H^{+} + 2\ e^{-}$$

$$F2 + 2\ e^{-} \rightarrow 2\ F^{-}$$

$$H_2 + F_2 \rightarrow 2\ H^{+} + 2\ F^{-}$$

And the ions combine to form hydrogen fluoride:

$$2\ H^{+} + 2\ F^{-} \rightarrow 2\ HF$$

The overall reaction is:

$$H2 + F2 \rightarrow 2\ HF$$

Metal Displacement

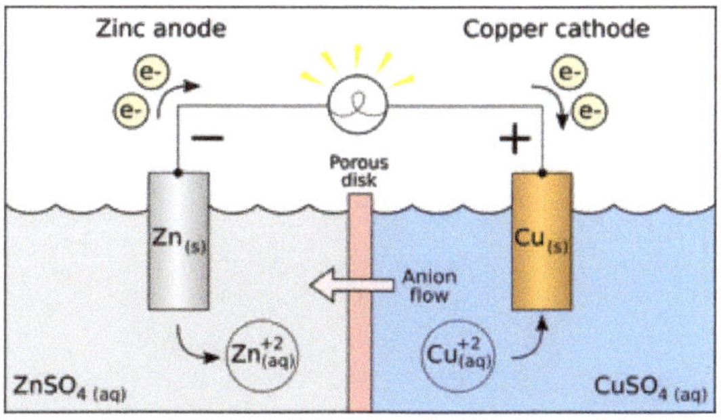

A redox reaction is the force behind an electrochemical cell like the Galvanic cell pictured. The battery is made out of a zinc electrode in a $ZnSO_4$ solution connected with a wire and a porous disk to a copper electrode in a $CuSO_4$ solution.

In this type of reaction, a metal atom in a compound (or in a solution) is replaced by an atom of another metal. For example, copper is deposited when zinc metal is placed in a copper(II) sulfate solution:

$$Zn(s) + CuSO_4(aq) \rightarrow ZnSO_4(aq) + Cu(s)$$

In the above reaction, zinc metal displaces the copper(II) ion from copper sulfate solution and thus liberates free copper metal.

The ionic equation for this reaction is:

$$Zn + Cu^{2+} \rightarrow Zn^{2+} + Cu$$

As two half-reactions, it is seen that the zinc is oxidized:

$$Zn \rightarrow Zn^{2+} + 2\,e^-$$

And the copper is reduced:

$$Cu^{2+} + 2\,e^- \rightarrow Cu$$

Other Examples

- The reduction of nitrate to nitrogen in the presence of an acid (denitrification):

$$2\,NO{-}3 + 10\,e^- + 12\,H^+ \rightarrow N_2 + 6\,H_2O$$

- The combustion of hydrocarbons, such as in an internal combustion engine, which produces water, carbon dioxide, some partially oxidized forms such as carbon monoxide, and heat energy. Complete oxidation of materials containing carbon produces carbon dioxide.
- In organic chemistry, the stepwise oxidation of a hydrocarbon by oxygen produces water and, successively, an alcohol, an aldehyde or a ketone, a carboxylic acid, and then a peroxide.

Corrosion and Rusting

Oxides, such as iron(III) oxide or rust, which consists of hydrated iron(III) oxides $Fe_2O_3 \cdot nH_2O$ and iron(III) oxide-hydroxide ($FeO(OH)$, $Fe(OH)_3$), form when oxygen combines with other elements

Iron rusting in pyrite cubes

- The term corrosion refers to the electrochemical oxidation of metals in reaction with an oxidant such as oxygen. Rusting, the formation of iron oxides, is a well-known example of electrochemical corrosion; it forms as a result of the oxidation of iron metal. Common rust often refers to iron(III) oxide, formed in the following chemical reaction:

  $4\ Fe + 3\ O_2 \rightarrow 2\ Fe_2O_3$

- The oxidation of iron(II) to iron(III) by hydrogen peroxide in the presence of an acid:

  $Fe^{2+} \rightarrow Fe^{3+} + e^-$

  $H_2O_2 + 2\ e^- \rightarrow 2\ OH^-$

  Overall equation:

  $2\ Fe^{2+} + H_2O_2 + 2\ H^+ \rightarrow 2\ Fe^{3+} + 2\ H_2O$

Redox Reactions in Industry

Cathodic protection is a technique used to control the corrosion of a metal surface by making it the cathode of an electrochemical cell. A simple method of protection connects protected metal to a more easily corroded "sacrificial anode" to act as the anode. The sacrificial metal instead of the protected metal, then, corrodes. A common application of cathodic protection is in galvanized steel, in which a sacrificial coating of zinc on steel parts protects them from rust.

The primary process of reducing ore at high temperature to produce metals is known as smelting.

Oxidation is used in a wide variety of industries such as in the production of cleaning products and oxidizing ammonia to produce nitric acid, which is used in most fertilizers.

Redox reactions are the foundation of electrochemical cells, which can generate electrical energy or support electrosynthesis.

The process of electroplating uses redox reactions to coat objects with a thin layer of a material, as in chrome-plated automotive parts, silver plating cutlery, and gold-plated jewelry.

The production of compact discs depends on a redox reaction, which coats the disc with a thin layer of metal film.

Redox Reactions in Biology

Many important biological processes involve redox reactions.

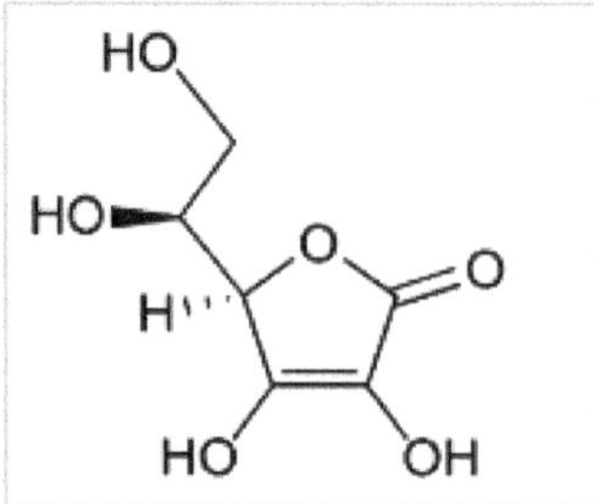

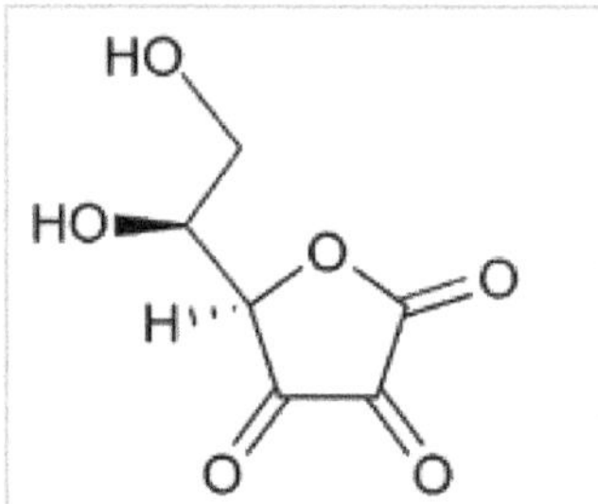

Top: ascorbic acid (reduced form of Vitamin C)
Bottom: dehydroascorbic acid (oxidized form of Vitamin C)

Cellular respiration, for instance, is the oxidation of glucose ($C_6H_{12}O_6$) to CO_2 and the reduction of oxygen to water. The summary equation for cell respiration is:

$$C_6H_{12}O_6 + 6\ O_2 \rightarrow 6\ CO_2 + 6\ H_2O$$

The process of cell respiration also depends heavily on the reduction of NAD^+ to NADH and the reverse reaction (the oxidation of NADH to NAD^+). Photosynthesis and cellular respiration are complementary, but photosynthesis is not the reverse of the redox reaction in cell respiration:

$$6\ CO_2 + 6\ H_2O + \text{light energy} \rightarrow C_6H_{12}O_6 + 6\ O_2$$

Biological energy is frequently stored and released by means of redox reactions. Photosynthesis involves the reduction of carbon dioxide into sugars and the oxidation of water into molecular oxygen. The reverse reaction, respiration, oxidizes sugars to produce carbon dioxide and water. As intermediate steps, the reduced carbon compounds are used to reduce nicotinamide adenine dinucleotide (NAD^+), which then contributes to the creation of a proton gradient, which drives the synthesis of adenosine triphosphate (ATP) and is maintained by the reduction of oxygen. In animal cells, mitochondria perform similar functions.

Free radical reactions are redox reactions that occur as a part of homeostasis and killing microorganisms, where an electron detaches from a molecule and then reattaches almost instantaneously. Free radicals are a part of redox molecules and can become harmful to the human body if they do not reattach to the redox molecule or an antioxidant. Unsatisfied free radicals can spur the mutation of cells they encounter and are, thus, causes of cancer.

The term redox state is often used to describe the balance of GSH/GSSG, NAD^+/NADH and $NADP^+$/NADPH in a biological system such as a cell or organ. The redox state is reflected in the balance of several sets of metabolites (e.g., lactate and pyruvate, beta-hydroxybutyrate, and acetoacetate), whose interconversion is dependent on these ratios. An abnormal redox state can develop in a variety of deleterious situations, such as hypoxia, shock, and sepsis. Redox mechanism also control some cellular processes. Redox proteins and their genes must be co-located for redox regulation according to the CoRR hypothesis for the function of DNA in mitochondria and chloroplasts.

Redox Cycling

A wide variety of aromatic compounds are enzymatically reduced to form free radicals that contain one more electron than their parent compounds. In general, the electron donor is any of a wide variety of flavoenzymes and their coenzymes. Once formed, these anion free radicals reduce molecular oxygen to superoxide, and regenerate the unchanged parent compound. The net reaction is the oxidation of the flavoenzyme's coenzymes and the reduction of molecular oxygen to form superoxide. This catalytic behavior has been described as futile cycle or redox cycling.

Examples of redox cycling-inducing molecules are the herbicide paraquat and other viologens and quinones such as menadione.

Redox Reactions in Geology

In geology, redox is important to both the formation of minerals and the mobilization of minerals, and is also important in some depositional environments. In general, the redox state of most rocks can be seen in the color of the rock. The rock forms in oxidizing conditions, giving it a red color. It is then "bleached" to a green—or sometimes white—form when a reducing fluid passes through the rock. The reduced fluid can also carry uranium-bearing minerals. Famous examples of redox conditions affecting geological processes include uranium deposits and Moqui marbles.

Mi Vida uranium mine, near Moab, Utah. The alternating red and white/green bands of sandstone correspond to oxidized and reduced conditions in groundwater redox chemistry.

Balancing Redox Reactions

Describing the overall electrochemical reaction for a redox process requires a *balancing* of the component half-reactions for oxidation and reduction. In general, for reactions in aqueous solution, this involves adding H^+, OH^-, H_2O, and electrons to compensate for the oxidation changes.

Acidic Media

In acidic media, H^+ ions and water are added to half-reactions to balance the overall reaction.

For instance, when manganese(II) reacts with sodium bismuthate:

Unbalanced reaction:	$Mn^{2+}(aq) + NaBiO_3(s) \rightarrow Bi^{3+}(aq) + MnO{-}4\ (aq)$
Oxidation:	$4\ H_2O(l) + Mn^{2+}(aq) \rightarrow MnO{-}4(aq) + 8\ H^+(aq) + 5\ e^-$
Reduction:	$2\ e^- + 6\ H^+ + BiO{-}3(s) \rightarrow Bi^{3+}(aq) + 3\ H_2O(l)$

The reaction is balanced by scaling the two half-cell reactions to involve the same number of electrons (multiplying the oxidation reaction by the number of electrons in the reduction step and vice versa):

$$8\ H_2O(l) + 2\ Mn^{2+}(aq) \rightarrow 2\ MnO{-}\ 4(aq) + 16\ H^+(aq) + 10\ e^-$$

$$10\ e^- + 30\ H^+ + 5\ BiO{-}\ 3(s) \rightarrow 5\ Bi^{3+}(aq) + 15\ H_2O(l)$$

Adding these two reactions eliminates the electrons terms and yields the balanced reaction:

$$14\ H^+(aq) + 2\ Mn^{2+}(aq) + 5\ NaBiO_3(s) \rightarrow 7\ H_2O(l) + 2\ MnO{-}$$
$$4(aq) + 5\ Bi^{3+}(aq) + 5\ Na+$$
$$(aq)$$

Basic Media

In basic media, OH^- ions and water are added to half reactions to balance the overall reaction.

For example, in the reaction between potassium permanganate and sodium sulfite:

Unbalanced reaction:	$KMnO_4 + Na_2SO_3 + H_2O \rightarrow MnO_2 + Na_2SO_4 + KOH$
Reduction:	$3\ e^- + 2\ H_2O + MnO{-}\ 4 \rightarrow MnO_2 + 4\ OH^-$
Oxidation:	$2\ OH^- + SO2{-}\ 3 \rightarrow SO2{-}\ 4 + H_2O + 2\ e^-$

Balancing the number of electrons in the two half-cell reactions gives:

$$6\ e^- + 4\ H_2O + 2\ MnO{-}\ 4 \rightarrow 2\ MnO_2 + 8\ OH^-$$

$$6\ OH^- + 3\ SO2-\ 3 \rightarrow 3\ SO2-4 + 3\ H_2O + 6\ e^-$$

Adding these two half-cell reactions together gives the balanced equation:

$$2\ KMnO_4 + 3\ Na_2SO_3 + H_2O \rightarrow 2\ MnO_2 + 3\ Na_2SO_4 + 2\ KOH$$

Memory Aids

The key terms involved in redox are often confusing to students. For example, an element that is oxidized loses electrons; however, that element is referred to as the reducing agent. Likewise, an element that is reduced gains electrons and is referred to as the oxidizing agent. Acronyms or mnemonics are commonly used to help remember the terminology:

- "OIL RIG" — oxidation is loss of electrons, reduction is gain of electrons.
- "LEO the lion says GER" — loss of electrons is oxidation, gain of electrons is reduction.
- "LEORA says GEROA" — loss of electrons is oxidation (reducing agent), gain of electrons is reduction (oxidizing agent).
- "RED CAT" and "AN OX", or "AnOx RedCat" ("an ox-red cat") — reduction occurs at the cathode and the anode is for oxidation.
- "RED CAT gains what AN OX loses" – reduction at the cathode gains (electrons) what anode oxidation loses (electrons).

Organic Redox Reaction

Organic reductions or organic oxidations or organic redox reactions are redox reactions that take place with organic compounds. In organic chemistry oxidations and reductions are different from ordinary redox reactions because many reactions carry the name but do not actually involve electron transfer in the electrochemical sense of the word. Instead the relevant criterion for organic oxidation is gain of oxygen and/or loss of hydrogen

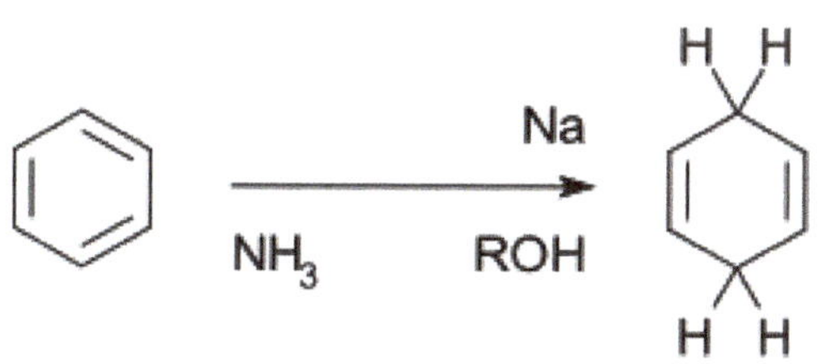

Organic redox reactions: The Birch reduction

Simple functional groups can be arranged in order of increasing oxidation state. The oxidation numbers are only an approximation:

oxidation number	compounds
−4	alkanes
−2	alkenes, alcohols, alkyl halides, amines
0	alkynes, ketones, aldehydes, geminal diols,
+2	carboxylic acids, amides, chloroform
+4	carbon dioxide, tetrachloromethane

When methane is oxidized to carbon dioxide its oxidation number changes from −4 to +4. Classical reductions include alkene reduction to alkanes and classical oxidations include oxidation of alcohols to aldehydes. In oxidations electrons are removed and the electron density of a molecule is reduced. In reductions electron density increases when electrons are added to the molecule. This terminology is always centered on the organic compound. For example, it is usual to refer to the reduction of a ketone by lithium aluminium hydride, but not to the oxidation of lithium aluminium hydride by a ketone. Many oxidations involve removal of hydrogen atoms from the organic molecule, and the reverse, reduction adds hydrogens to an organic molecule.

Many reactions classified as reductions also appear in other classes. For instance conversion of the ketone to an alcohol by lithium aluminium hydride can be considered a reduction but the hydride is also a good nucleophile in nucleophilic substitution. Many redox reactions in organic chemistry have coupling reaction reaction mechanism involving free radical intermediates. True organic redox chemistry can be found in electrochemical organic synthesis or electrosynthesis. Examples of organic reactions that can take place in an electrochemical cell are the Kolbe electrolysis.

In disproportionation reactions the reactant is both oxidised and reduced in the same chemical reaction forming two separate compounds.

Asymmetric catalytic reductions and asymmetric catalytic oxidations are important in asymmetric synthesis.

Organic Oxidations

Most oxidations are conducted with air or oxygen. These oxidation include routes to chemical compounds, remediation of pollutants, and combustion. Several reaction mechanisms exist for organic oxidations:

- Single electron transfer
- Oxidations through ester intermediates with chromic acid or manganese dioxide
- Hydrogen atom transfer as in Free radical halogenation

- Oxidation involving ozone in ozonolysis or peroxides (e.g. Peroxy acids)
- Oxidations involving an elimination reaction mechanism such as the Swern oxidation, the Kornblum oxidation and with reagents such as IBX acid and Dess-Martin periodinane.
- Oxidation by nitroxide radicals Fremy's salt or TEMPO

Organic Reductions

Several reaction mechanisms exist for organic reductions:

- Direct electron transfers (e.g. Birch reduction).
- Hydride transfer in reductions with for example lithium aluminium hydride or a hydride shift as in the Meerwein-Ponndorf-Verley reduction
- Hydrogenations using a variety of catalysts (e.g. Raney nickel or Platinum dioxide) or specific reductions (e.g. named reactions such as Rosenmund reduction).
- Disproportionation reaction such as the Cannizzaro reaction

Reductions that do not fit in any reduction reaction mechanism and in which just the change in oxidation state is reflected include the Wolff-Kishner reaction.

Electrochlorination

Electrochlorination is the process of producing hypochlorite by running an electric current through salt water. This is used to disinfect water and make it safe for human use, such as for drinking water or swimming pools.

Process

The processes involved in electrochlorination are actually rather simple. The main idea involved is the electrolysis of water to produce a chlorinated solution. This happens when saltwater is inserted into electrolyzer cells. The first step is removing the solid excess from the saltwater. Next, as the saltwater runs it is streamed through a channel of decreasing thickness. One side of the channel is a cathode, the other is an anode. As the water flows through the anode/cathode channel, a low voltage DC currents applied. When this happens, the electrolysis is triggered and sodium hypochlorite is instantly produced as well as hydrogen gas (H_2). The hydrogen rich sodium hypochlorite then travels to a tank that removes the hydrogen gas. The dehydrogenization mechanism varies from device to device but the process is generally the same. After hydrogen has been removed from the solution, it is stored in a tank as the finished product. No chemicals other than ordinary salt, or sodium chloride (NaCl), are used throughout the entirety of the process. Although the actual chemical processes involved are complex, they can be simply represented by the following equation:

$$NaCl + H_2O + ENERGY \rightarrow NaOCl + H_2$$

In words this reads, energy is added to sodium chloride (table salt) in water, resulting in sodium hypochlorite and hydrogen gas.

Seawater

On occasion, companies will want to use seawater for this process due to its extreme low obtaining cost. The seawater used is usually brackish water or brine (i.e. a solution with >0.5% salinity). In these cases, random chemicals may be introduced into the system. These random chemicals pose no threat to the system however due to the simplicity of the process. The low voltage DC current that is applied will do only what it is intended to do, i.e. electrochlorination. The excess random chemicals are left untouched and can be easily discarded.

Products

The product of this process, sodium hypochlorite, contains 0.7% to 1% chlorine. Anything below the concentration of 1% chlorine is considered a non-hazardous chemical although still a very effective disinfectant. In addition, the sodium hypochlorite produced is in the pH range of 6-7.5. This means that the chemical is relatively neutral in regards to acidity or baseness. Also, at that pH range, the sodium hypochlorite is extremely stable and the electrochlorination extremely effective.

Applications

Drinking Water

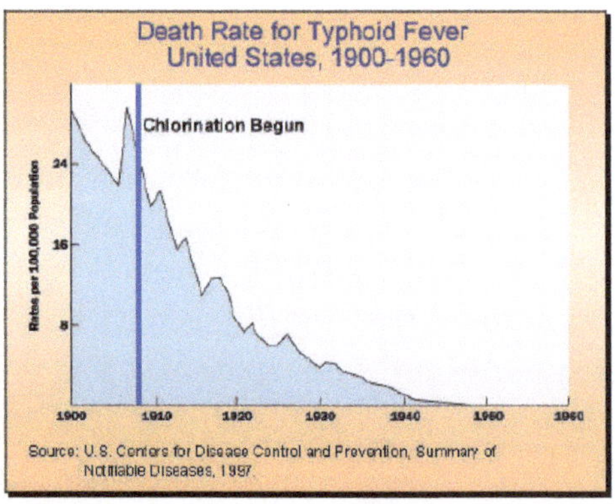

Electrochlorination is the next step in the evolution of this process. Electrochlorination chlorinates drinking water and does so in an Eco-friendly way. It does not harm the environment in any measurable negative way. Unlike other chlorination techniques, electrochlorination generates no sludge or by-products. It also is safer for the operators of the chlorinators as there is no handling of chlorine gas, which is highly toxic and corrosive.

Local water treatment plants have begun to evolve their technology over the years as health threats due to water contamination become more known including cholera, typhoid, and dysentery. to combat these threats, water treatment plants began to implement chlorination. Chlorination virtually wiped out both the spread and initial contamination of these diseases, and did so in a way that earned it the title of "the most significant public health advance of the millennium" from Life Magazine.

Swimming Pools

When a swimmer enters a pool, they add up to one billion organisms to the water. Chlorination kills all organisms harmful to swimmers such as those that cause ear infections and athlete's foot. The advantages of electrochlorination in this process are as follows:

- Not irritating to skin or soft tissue.
- Active in small concentrations.
- Longer lifespan of chemical and therefore the less replacement necessary.
- Easily measurable.

Electrochlorination chlorinates water in an Eco-friendly manner. Unlike other chlorination techniques, electrochlorination generates no sludge or by-products. It also is safer for the operators of the chlorinators as there is no handling of chlorine gas, which is highly toxic and corrosive.

Electrolysis

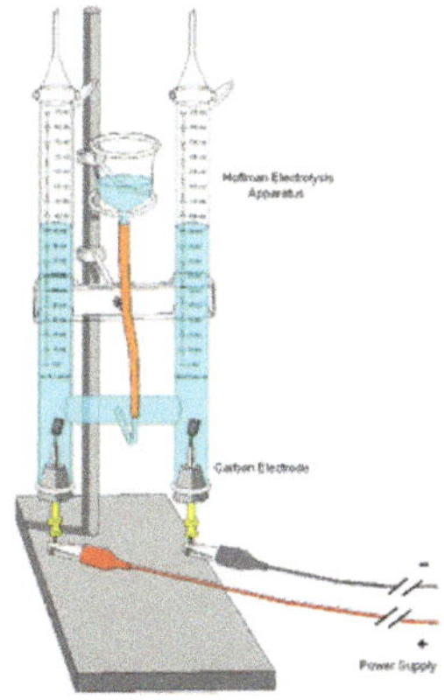

Illustration of an electrolysis apparatus used in a school laboratory.

In chemistry and manufacturing, electrolysis is a technique that uses a direct electric current (DC) to drive an otherwise non-spontaneous chemical reaction. Electrolysis is commercially important as a stage in the separation of elements from naturally occurring sources such as ores using an electrolytic cell. The voltage that is needed for electrolysis to occur is called the decomposition potential.

History

The word "electrolysis" was introduced by Michael Faraday in the 19th century, on the suggestion of the Rev. Which since the 17th century was associated with electric phenomena, and λύσις [lýsis] meaning "dissolution". Nevertheless, electrolysis, as a tool to study chemical reactions and obtain pure elements, precedes the coinage of the term and formal description by Faraday.

- 1785 – Martinus van Marum's electrostatic generator was used to reduce tin, zinc, and antimony from their salts using electrolysis.
- 1800 – William Nicholson and Anthony Carlisle (view also Johann Ritter), decomposed water into hydrogen and oxygen.
- 1808 – Potassium (1807), sodium, barium, calcium and magnesium were discovered by Sir Humphry Davy using electrolysis.
- 1821 – Lithium was discovered by William Thomas Brande who obtained it by electrolysis of lithium oxide.
- 1833 – Michael Faraday develops his two laws of electrolysis, and provides a mathematical explanation of his laws.
- 1875 – Paul Émile Lecoq de Boisbaudran discovered gallium using electrolysis.
- 1886 – Fluorine was discovered by Henri Moissan using electrolysis.
- 1886 – Hall–Héroult process developed for making aluminium
- 1890 – Castner–Kellner process developed for making sodium hydroxide

Overview

Electrolysis is the passing of a direct electric current through an ionic substance that is either molten or dissolved in a suitable solvent, producing chemical reactions at the electrodes and separation of materials.

The main components required to achieve electrolysis are:

- An electrolyte: a substance, frequently an ion-conducting polymer that contains free ions, which carry electric current in the electrolyte. If the ions are not mobile, as in a solid salt then electrolysis cannot occur.
- A direct current (DC) electrical supply: provides the energy necessary to create or discharge the ions in the electrolyte. Electric current is carried by electrons in the external circuit.

- Two electrodes: electrical conductors that provide the physical interface between the electrolyte and the electrical circuit that provides the energy.

Electrodes of metal, graphite and semiconductor material are widely used. Choice of suitable electrode depends on chemical reactivity between the electrode and electrolyte and manufacturing cost.

Process of Electrolysis

The key process of electrolysis is the interchange of atoms and ions by the removal or addition of electrons from the external circuit. The desired products of electrolysis are often in a different physical state from the electrolyte and can be removed by some physical processes. For example, in the electrolysis of brine to produce hydrogen and chlorine, the products are gaseous. These gaseous products bubble from the electrolyte and are collected.

$$2\,NaCl + 2\,H_2O \rightarrow 2\,NaOH + H_2 + Cl_2$$

A liquid containing mobile ions (electrolyte) is produced by:

- Solvation or reaction of an ionic compound with a solvent (such as water) to produce mobile ions
- An ionic compound is fused by heating

An electrical potential is applied across a pair of electrodes immersed in the electrolyte.

Each electrode attracts ions that are of the opposite charge. Positively charged ions (cations) move towards the electron-providing (negative) cathode. Negatively charged ions (anions) move towards the electron-extracting (positive) anode.

In this process electrons are either absorbed or released. Neutral atoms gain or lose electrons and become charged ions that then pass into the electrolyte. The formation of uncharged atoms from ions is called discharging. When an ion gains or loses enough electrons to become uncharged (neutral) atoms, the newly formed atoms separate from the electrolyte. Positive metal ions like Cu^{++} deposit onto the cathode in a layer. The terms for this are electroplating, electrowinning, and electrorefining. When an ion gains or loses electrons without becoming neutral, its electronic charge is altered in the process. In chemistry, the loss of electrons is called Oxidation, while electron gain is called reduction.

Oxidation and Reduction at the Electrodes

Oxidation of ions or neutral molecules occurs at the anode. For example, it is possible to oxidize ferrous ions to ferric ions at the anode:

$$Fe2+\ (aq) \rightarrow Fe3+\ (aq) + e^-$$

Reduction of ions or neutral molecules occurs at the cathode.

It is possible to reduce ferricyanide ions to ferrocyanide ions at the cathode:

$$Fe(CN)3\text{- }6 + e^- \rightarrow Fe(CN)4\text{- }6$$

Neutral molecules can also react at either of the electrodes. For example: p-Benzoquinone can be reduced to hydroquinone at the cathode:

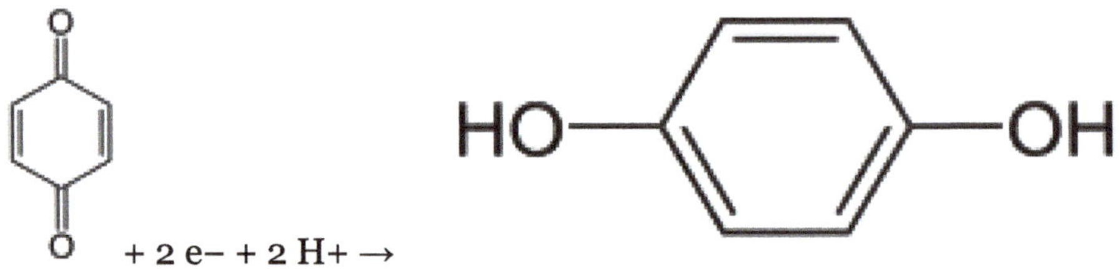

In the last example, H^+ ions (hydrogen ions) also take part in the reaction, and are provided by an acid in the solution, or by the solvent itself (water, methanol etc.). Electrolysis reactions involving H^+ ions are fairly common in acidic solutions. In aqueous alkaline solutions, reactions involving OH^- (hydroxide ions) are common.

Sometimes the solvents themselves (usually water) are oxidized or reduced at the electrodes. It is even possible to have electrolysis involving gases. Such as when using a Gas diffusion electrode.

Energy Changes During Electrolysis

The amount of electrical energy that must be added equals the change in Gibbs free energy of the reaction plus the losses in the system. The losses can (in theory) be arbitrarily close to zero, so the maximum thermodynamic efficiency equals the enthalpy change divided by the free energy change of the reaction. In most cases, the electric input is larger than the enthalpy change of the reaction, so some energy is released in the form of heat. In some cases, for instance, in the electrolysis of steam into hydrogen and oxygen at high temperature, the opposite is true and heat energy is absorbed. This heat is absorbed from the surroundings, and the heating value of the produced hydrogen is higher than the electric input.

Related Techniques

The following techniques are related to electrolysis:

- Electrochemical cells, including the hydrogen fuel cell, use differences in Standard electrode potential to generate an electrical potential that provides useful power. Though related via the interaction of ions and electrodes, electrolysis and the operation of electrochemical cells are quite distinct. However, a chemical cell should *not* be seen as performing *electrolysis in reverse*.

Faraday's Laws of Electrolysis

First Law of Electrolysis

In 1832, Michael Faraday reported that the quantity of elements separated by passing an electric current through a molten or dissolved salt is proportional to the quantity of electric charge passed through the circuit. This became the basis of the first law of electrolysis:

$$m = k \cdot q$$

or

$$m = eQ$$

where; e is known as electrochemical equivalent of the metal deposited or of the gas liberated at the electrode.

Second Law of Electrolysis

Faraday discovered that when the same amount of current is passed through different electrolytes/elements connected in series, the mass of substance liberated/deposited at the electrodes is directly proportional to their equivalent weight.

Industrial Uses

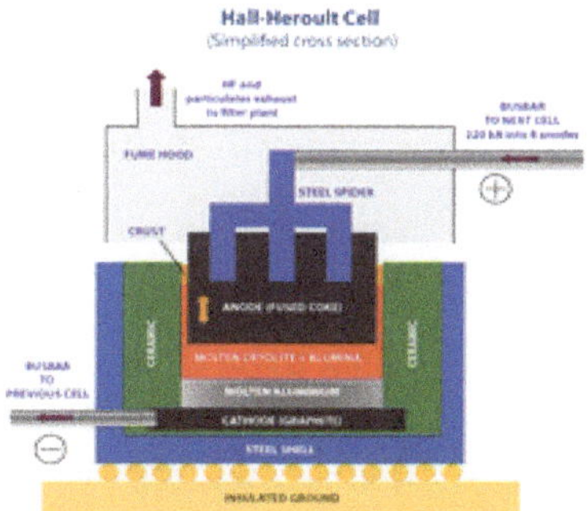

Hall-Heroult process for producing aluminium

- Electrometallurgy is the process of reduction of metals from metallic compounds to obtain the pure form of metal using electrolysis. aluminium, lithium, sodium, potassium, magnesium, calcium, and in some cases copper, are produced in this way.
- Production of chlorine and sodium hydroxide
- Production of sodium chlorate and potassium chlorate
- Production of perfluorinated organic compounds such as trifluoroacetic acid by the process of electrofluorination

- Production of electrolytic copper as a cathode, from refined copper of lower purity as an anode.

Electrolysis has many other uses:

- Production of oxygen for spacecraft and nuclear submarines.
- Production of hydrogen for fuel, using a cheap source of electrical energy.

Electrolysis is also used in the cleaning and preservation of old artifacts. Because the process separates the non-metallic particles from the metallic ones, it is very useful for cleaning a wide variety of metallic objects, from old coins to even larger objects including rusted cast iron cylinder blocks and heads when rebuilding automobile engines. Rust removal from small iron or steel objects by electrolysis can be done in a home workshop using simple materials such as a plastic bucket, tap water, lengths of rebar, washing soda, baling wire, and a battery charger.

Manufacturing Processes

In manufacturing, electrolysis can be used for:

- Electroplating, where a thin film of metal is deposited over a substrate material. Electroplating is used in many industries for either functional or decorative purposes, as in vehicle bodies and nickel coins.
- Electrochemical Machining (ECM), where an electrolytic cathode is used as a shaped tool for removing material by anodic oxidation from a workpiece. ECM is often used as technique for deburring or for etching metal surfaces like tools or knives with a permanent mark or logo.

Competing Half-reactions in Solution Electrolysis

Using a cell containing inert platinum electrodes, electrolysis of aqueous solutions of some salts leads to reduction of the cations (e.g., metal deposition with, e.g., zinc salts) and oxidation of the anions (e.g. evolution of bromine with bromides). However, with salts of some metals (e.g. sodium) hydrogen is evolved at the cathode, and for salts containing some anions (e.g. sulfate SO_4^{2-}) oxygen is evolved at the anode. In both cases this is due to water being reduced to form hydrogen or oxidized to form oxygen. In principle the voltage required to electrolyze a salt solution can be derived from the standard electrode potential for the reactions at the anode and cathode. The standard electrode potential is directly related to the Gibbs free energy, ΔG, for the reactions at each electrode and refers to an electrode with no current flowing. An extract from the table of standard electrode potentials is shown below.

Half-reaction	*E*° (V)	Ref.
$Na^+ + e^- \rightleftharpoons Na(s)$	−2.71	

$Zn^{2+} + 2e^- \rightleftharpoons Zn(s)$	−0.7618	
$2H^+ + 2e^- \rightleftharpoons H_2(g)$	≡ 0	
$Br_2(aq) + 2e^- \rightleftharpoons 2Br^-$	+1.0873	
$O_2(g) + 4H^+ + 4e^- \rightleftharpoons 2H_2O$	+1.23	
$Cl_2(g) + 2e^- \rightleftharpoons 2Cl^-$	+1.36	
$S2O^{2-}$ 8 + $2e^- \rightleftharpoons$ 2SO2−4	+2.07	

In terms of electrolysis, this table should be interpreted as follows:

- Oxidized species (often a cation) with a more negative cell potential are more difficult to reduce than oxidized species with a more positive cell potential. For example, it is more difficult to reduce a sodium ion to a sodium metal than it is to reduce a zinc ion to a zinc metal.
- Reduced species (often an anion) with a more positive cell potential are more difficult to oxidize than reduced species with a more negative cell potential. For example, it is more difficult to oxidize sulfate anions than it is to oxidize bromide anions.

Using the Nernst equation the electrode potential can be calculated for a specific concentration of ions, temperature and the number of electrons involved. For pure water (pH 7):

- the electrode potential for the reduction producing hydrogen is −0.41 V
- the electrode potential for the oxidation producing oxygen is +0.82 V.

Comparable figures calculated in a similar way, for 1M zinc bromide, $ZnBr_2$, are −0.76 V for the reduction to Zn metal and +1.10 V for the oxidation producing bromine. The conclusion from these figures is that hydrogen should be produced at the cathode and oxygen at the anode from the electrolysis of water—which is at variance with the experimental observation that zinc metal is deposited and bromine is produced. The explanation is that these calculated potentials only indicate the thermodynamically preferred reaction. In practice many other factors have to be taken into account such as the kinetics of some of the reaction steps involved. These factors together mean that a higher potential is required for the reduction and oxidation of water than predicted, and these are termed overpotentials. Experimentally it is known that overpotentials depend on the design of the cell and the nature of the electrodes.

For the electrolysis of a neutral (pH 7) sodium chloride solution, the reduction of sodium ion is thermodynamically very difficult and water is reduced evolving hydrogen leaving hydroxide ions in solution. At the anode the oxidation of chlorine is observed rather than the oxidation of water since the overpotential for the oxidation of chloride to chlorine is lower than the overpotential for the oxidation of water to oxygen. The hydroxide ions and dissolved chlorine gas react further to form hypochlorous acid. The aqueous solutions resulting from this process is called electrolyzed water and is used as a disinfectant and cleaning agent.

Research Trends

Electrolysis of Carbon Dioxide

The electrochemical reduction or electrocatalytic conversion of CO_2 can produce value-added chemicals such methane, ethylene, ethane, etc. The electrolysis of carbon dioxide gives formate or carbon monoxide, but sometimes more elaborate organic compounds such as ethylene. This technology is under research as a carbon-neutral route to organic compounds.

Electrolysis of Water

Electrolysis of water produces hydrogen.

$$2\ H_2O(l) \rightarrow 2\ H_2(g) + O_2(g);\ E_0 = +1.229\ V$$

The energy efficiency of water electrolysis varies widely. The efficiency of an electrolyser is a measure of the enthalpy contained in the hydrogen (to undergo combustion with oxygen, or some other later reaction), compared with the input electrical energy. Heat/enthalpy values for hydrogen are well published in science and engineering texts, as 144 MJ/kg. Note that fuel cells (not electrolysers) cannot utilise this full amount of heat/enthalpy, which has led to some confusion when calculating efficiency values for both types of technology. In the reaction, some energy is lost as heat. Some reports quote efficiencies between 50% and 70% for alkaline electrolysers; however, much higher practical efficiencies are available with the use of PEM (Polymer Electrolyte Membrane electrolysis) and catalytic technology, such as 95% efficiency.

NREL estimated that 1 kg of hydrogen (roughly equivalent to 3 kg, or 4 L, of petroleum in energy terms) could be produced by wind powered electrolysis for between $5.55 in the near term and $2.27 in the long term.

About 4% of hydrogen gas produced worldwide is generated by electrolysis, and normally used onsite. Hydrogen is used for the creation of ammonia for fertilizer via the Haber process, and converting heavy petroleum sources to lighter fractions via hydrocracking.

Carbon/hydrocarbon Assisted Water Electrolysis

Recently, to reduce the energy input, the utilization of carbon (coal), alcohols (hydrocarbon solution), and organic solution (glycerol, formic acid, ethylene glycol, etc.) with co-electrolysis of water has been proposed as a viable option. The carbon/hydrocarbon assisted electrolysis for hydrogen generation would performs this operation in a single electrochemical reactor. This system energy balance can be required only around 40% electric input with 60% coming from the chemical energy of carbon or hydrocarbon.

Electrocrystallization

A specialized application of electrolysis involves the growth of conductive crystals on one of the electrodes from oxidized or reduced species that are generated in situ. The technique has been used to obtain single crystals of low-dimensional electrical conductors, such as charge-transfer salts.

History

Scientific pioneers of electrolysis include:

- Antoine Lavoisier
- Robert Bunsen
- Humphry Davy
- Michael Faraday
- Paul Héroult
- Svante Arrhenius
- Adolph Wilhelm Hermann Kolbe
- William Nicholson
- Joseph Louis Gay-Lussac
- Alexander von Humboldt
- Johann Wilhelm Hittorf
- Kai Grjotheim-->

Pioneers of batteries:

- Alessandro Volta
- Gaston Planté

References

- Todd, Robert H.; Dell K. Allen; Leo Alting (1994). "Surface Coating". Manufacturing Processes Reference Guide. Industrial Press Inc. pp. 454–458. ISBN 0-8311-3049-0.
- Thomas, John Meurig (1991). Michael Faraday and the Royal Institution: The Genius of Man and Place. Bristol: Hilger. p. 51. ISBN 0750301457.
- Bard, Allan; Inzelt, Gyorgy; Scholz, Fritz, eds. (2012). "Haring-Blum Cell". Electrochemical Dictionary. Springer. p. 444. doi:10.1007/978-3-642-29551-5_8. ISBN 978-3-642-29551-5.
- Wendt, Hartmut; Kreyse, Gerhard (1999). Electrochemical Engineering: Science and Technology in Chemical and Other Industries. Springer. p. 122. ISBN 3540643869.

- Hudlický, Miloš (1996). Reductions in Organic Chemistry. Washington, D.C.: American Chemical Society. p. 429. ISBN 0-8412-3344-6.
- Robertson, William (2010). More Chemistry Basics. National Science Teachers Association. p. 82. ISBN 978-1-936137-74-9.
- Phillips, John; Strozak, Victor; Wistrom, Cheryl (2000). Chemistry: Concepts and Applications. Glencoe McGraw-Hill. p. 558. ISBN 978-0-02-828210-7.
- Rodgers, Glen (2012). Descriptive Inorganic, Coordination, and Solid-State Chemistry. Brooks/ Cole, Cengage Learning. p. 330. ISBN 978-0-8400-6846-0.
- March Jerry; (1985). Advanced Organic Chemistry reactions, mechanisms and structure (3rd ed.). New York: John Wiley & Sons, inc. ISBN 0-471-85472-7
- R. J. D. Tilley (2004). Understanding solids: the science of materials. John Wiley and Sons. pp. 281–. ISBN 978-0-470-85276-7. Retrieved 22 October 2011.

6

Theory and Laws Related to Electrochemistry

The theories and laws related to electrochemistry are Debye–Hückel theory, Paschen's law and Faraday's law of electrolysis. Debye-Hückel theory explains the departures from ideality in solutions of electrolytes and plasmas. The topics discussed in the text are of great importance to broaden the existing knowledge on electrochemistry.

Debye–Hückel Theory

The Debye–Hückel theory was proposed by Peter Debye and Erich Hückel as a theoretical explanation for departures from ideality in solutions of electrolytes and plasmas. It is a linearized Poisson–Boltzmann model, which assumes an extremely simplified model of the electrolyte solution but nevertheless gave accurate predictions of mean activity coefficients for ions in dilute solution. The Debye–Hückel equation provides a starting point for modern treatments of non-ideality of electrolyte solutions.

Overview

In the chemistry of electrolyte solutions, an ideal solution is a solution whose colligative properties are proportional to the concentration of the solute. Real solutions show departures from this kind of ideality at all but the very lowest concentrations. In order to accommodate these effects in the thermodynamics of solutions, the concept of activity was introduced: the properties are then proportional to the activities of the ions. Activity, a, is proportional to concentration, c. The proportionality constant is known as an activity coefficient, γ.

$$a = \gamma c$$

In an ideal electrolyte solution the activity coefficients of all the ions are equal to one. Non-ideality arises principally (but not exclusively) because ions of opposite charge attract each other due to electrostatic forces, while ions of the same charge repel each other. In consequence ions are not randomly distributed throughout the solution, as they would be in an ideal solution.

Activity coefficients of single ions cannot be measured experimentally because an electrolyte solution must contain both positively charged ions and negatively charged ions. Instead, a mean activity coefficient, $\gamma_{\pm}$ is defined. For example, with the electrolyte NaCl

$$\gamma_{\pm} = \left(\gamma_{Na^+}\gamma_{Cl^-}\right)^{1/2}$$

In general, the mean activity coefficient of a fully dissociated electrolyte of formula A_nB_m is given by

$$\gamma_{\pm} = \left(\gamma_A{}^n \gamma_B{}^m\right)^{1/(n+m)}$$

Activity coefficients are themselves functions of concentration as the amount of inter-ionic interaction increases as the concentration of the electrolyte increases. Debye and Hückel developed a theory with which single ion activity coefficients could be calculated. By calculating the mean activity coefficients from them the theory could be tested against experimental data. It was found to give excellent agreement for "dilute" solutions.

The Model

A description of Debye–Hückel theory includes a very detailed discussion of the assumptions and their limitations as well as the mathematical development and applications.

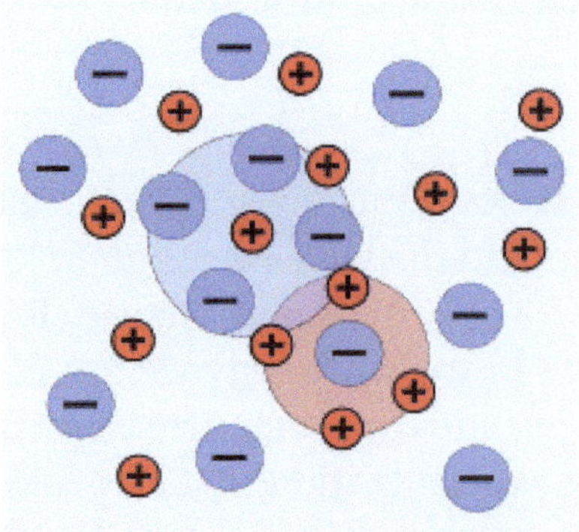

An idealized representation of a solution of a 1:1 electrolyte

A snapshot of a 2-dimensional section of an idealized electrolyte solution is shown in the picture. The ions are shown as spheres with unit electrical charge. The solvent (pale blue) is shown as a uniform medium, without structure. On average, each ion is surrounded more closely by ions of opposite charge than by ions of like charge. These concepts were developed into a quantitative theory involving ions of charge z_1e^+ and z_2e^-, where z can be any integer. The principal assumption is that departure from ideality is due to electrostatic interactions between ions, mediated by Coulomb's law: the force of interaction between two electric charges, separated by a distance, r in a medium of relative permittivity ε_r is given by

$$\text{force} = \frac{z_1 z_2 e^2}{4\pi\epsilon_0\epsilon_r r^2}$$

It is also assumed that

- The solute is completely dissociated; it is a strong electrolyte.
- Ions are spherical and are not polarized by the surrounding electric field. Solvation of ions is ignored except insofar as it determines the effective sizes of the ions.
- The solvent plays no role other than providing a medium of constant relative permittivity (dielectric constant).
- There is no electrostriction.
- Individual ions surrounding a "central" ion can be represented by a statistically averaged cloud of continuous charge density, with a minimum distance of closest approach.

The last assumption means that each cation is surrounded by a spherically symmetric cloud of other ions. The cloud has a net negative charge. Similarly each anion is surrounded by a cloud with net positive charge.

Mathematical Development

The deviation from ideality is taken to be a function of the potential energy resulting from the electrostatic interactions between ions and their surrounding clouds. To calculate this energy two steps are needed.

The first step is to specify the electrostatic potential for ion j by means of Poisson's equation

$$\nabla^2\psi_j(r) = -\frac{1}{\epsilon_0\epsilon_r}\rho_j(r)$$

$\psi(r)$ is the total potential at a distance, r, from the central ion and $\rho(r)$ is the averaged charge density of the surrounding cloud at that distance. To apply this formula it is essential that the cloud has spherical symmetry, that is, the charge density is a function only of distance from the central ion as this allows the Poisson equation to be cast in terms of spherical coordinates with no angular dependence.

The second step is to calculate the charge density by means of a Maxwell–Boltzmann distribution.

$$n_i' = n_i \exp\left(\frac{-z_i e\psi_j(r)}{kT}\right)$$

This distribution also depends on the potential $\psi(r)$ and this introduces a serious difficulty in terms of the superposition principle. Nevertheless, the two equations can be combined to produce the Poisson–Boltzmann equation.

$$\nabla^2\psi_j(r) = -\frac{1}{\epsilon_0\epsilon_r}\sum_i\left\{n_i(z_ie)\exp\left(\frac{-z_ie\psi_j(r)}{kT}\right)\right\}$$

Solution of this equation is far from straightforward. Debye and Hückel expanded the exponential as a truncated Taylor series to first order. The zeroth order term vanishes because the solution is on average electrically neutral (so that $\sum n_i z_i = 0$), which leaves us with only the first order term. The result has the form of the Helmholtz equation

$$\nabla^2\psi_j(r) = \kappa^2\psi_j(r) \qquad \text{with} \qquad \kappa^2 = \frac{e^2}{\epsilon_0\epsilon_r kT}\sum_i n_i z_i^2 ,$$

which has an analytical solution. This equation applies to electrolytes with equal numbers of ions of each charge. Nonsymmetrical electrolytes require another term with ψ^2. For symmetrical electrolytes, this reduces to the modified spherical Bessel equation

$$(\partial_r^2 + \frac{2}{r}\partial_r - \kappa^2)\psi_j = 0 \qquad \text{with solutions} \qquad \psi_j(r) = A'\frac{e^{-\kappa r}}{r} + A''\frac{e^{\kappa r}}{r}$$

The coefficients A' and A'' are fixed by the boundary conditions. As $r \to \infty$, ψ must not diverge, so $A'' = 0$. At $r = a_0$, which is the distance of the closest approach of ions, the force exerted by the charge should be balanced by the force of other ions, imposing $\partial_r\psi_j(a_0) = -z_je/(4\pi\varepsilon_0\varepsilon_r a_0^2)$, from which A' is found, yielding

$$\psi_j(r) = \frac{z_je}{4\pi\varepsilon_0\varepsilon_r}\frac{e^{\kappa a_0}}{1+\kappa a_0}\frac{e^{-\kappa r}}{r}$$

The electrostatic potential energy, u_j, of the ion at $r = 0$ is

$$u_j = z_je\left(\psi_j(a_0) - \frac{z_je}{4\pi\varepsilon_0\varepsilon_r}\frac{1}{a_0}\right) = -\frac{z_j^2e^2}{4\pi\varepsilon_0\varepsilon_r}\frac{\kappa}{1+\kappa a_0}$$

This is the potential energy of a single ion in a solution. The multiple-charge generalization from electrostatics gives an expression for the potential energy of the entire solution. The mean activity coefficient is given by the logarithm of this quantity as follows

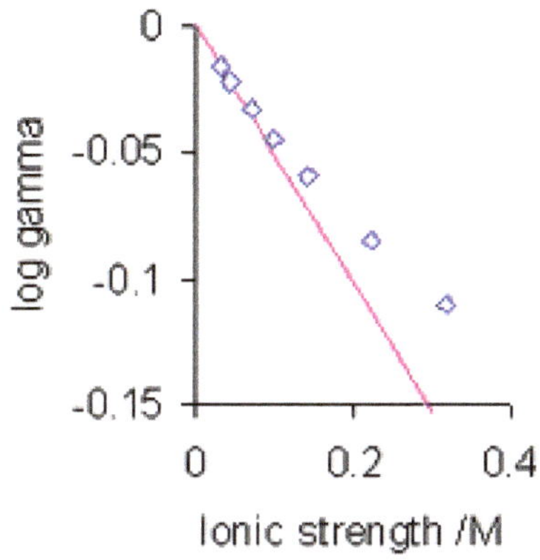

Experimental $\log \gamma_{\pm}$ values for KBr at 25°C (points) and Debye–Hückel limiting law (coloured line)

$$\log_{10} \gamma_{\pm} = -Az_j^2 \frac{\sqrt{I}}{1 + Ba_0\sqrt{I}}$$

$$A = \frac{e^2 B}{2.303 \times 8\pi\epsilon_0\epsilon_r kT}$$

$$B = \left(\frac{2e^2 N}{\epsilon_0\epsilon_r kT}\right)^{1/2}$$

where I is the ionic strength and a_0 is a parameter that represents the distance of closest approach of ions. For aqueous solutions at 25 °C A = 0.51 $mol^{-1/2}dm^{3/2}$ and B = 3.29 $nm^{-1}mol^{-1/2}dm^{3/2}$

The most significant aspect of this result is the prediction that the mean activity coefficient is a function of *ionic strength* rather than the electrolyte concentration. For very low values of the ionic strength the value of the denominator in the expression above becomes nearly equal to one. In this situation the mean activity coefficient is proportional to the square root of the ionic strength. This is known as the Debye–Hückel limiting law.

Limitations and Extensions

The equation for $\log \gamma_{\pm}$ gives satisfactory agreement with experimental measurements for low electrolyte concentrations, typically less than 10^{-3} mol dm^{-3}. Deviations from the theory occur at higher concentrations and with electrolytes that produce ions of higher charges, particularly unsymmetrical electrolytes. Essentially these deviations occur because the model is hopelessly oversimplified, so there is little to be gained making small adjustments to the model. The individual assumptions can be challenged in turn.

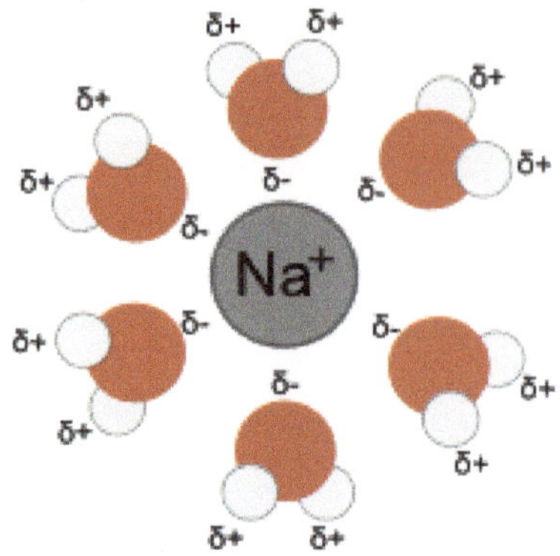

The first solvation shell of a sodium ion dissolved in water. The oxygen atoms are arranged at the vertices of an octahedron with the sodium ion at its centre

- Complete dissociation. Ion-association may take place, particularly with ions of higher charge. This was followed up in detail by Niels Bjerrum. The Bjerrum length is the separation at which the electrostatic interaction between two ions is comparable in magnitude to kT.

- Weak electrolytes. A weak electrolyte is one that is not fully dissociated. As such it has a dissociation constant. The dissociation constant can be used to calculate the extent of dissociation and hence, make the necessary correction needed to calculate activity coefficients.

- Ions are spherical, not point charges and are not polarized. Many ions such as the nitrate ion, NO_3^-, are manifestly not spherical. Polyatomic ions are also polarizable.

- Role of the solvent. The solvent is not a structureless medium but is made up of molecules. The water molecules in aqueous solution are both dipolar and polarizable. Both cations and anions have a strong primary solvation shell and a weaker secondary solvation shell. Ion-solvent interactions are ignored in Debye–Hückel theory.

Moreover, ionic radius is assumed to be negligible, but at higher concentrations, the ionic radius becomes comparable to the radius of the ionic atmosphere. Most extensions to Debye–Hückel theory are empirical in nature. They usually allow the Debye–Hückel equation to be followed at low concentration and add further terms in some power of the ionic strength to fit experimental observations. The main extensions are the Davies equation, Pitzer equations and Specific ion interaction theory.

Electrolytes Mixtures

The theory can be applied also to dilute solutions of mixed electrolytes. Freezing point depression measurements has been used to this purpose.

Conductivity

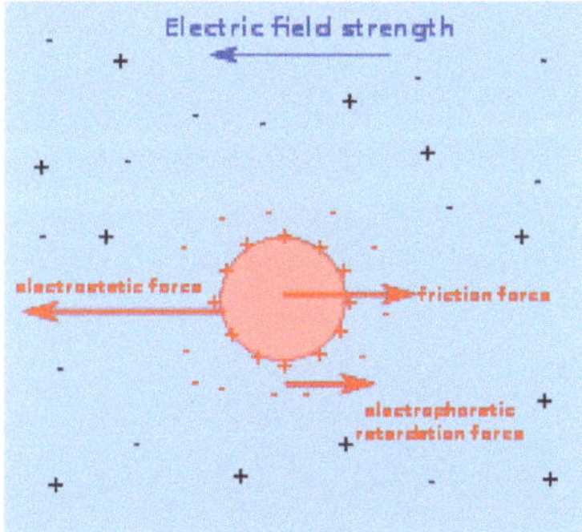

Forces affecting a charged ion or cloud moving in an electric field

The treatment given so far is for a system not subject to an external electric field. When conductivity is measured the system is subject to an oscillating external field due to the application of an AC voltage to electrodes immersed in the solution. Debye and Hückel modified their theory in 1926 and their theory was further modified by Lars Onsager in 1927. All the postulates of the original theory were retained. In addition it was assumed that the electric field causes the charge cloud to be distorted away from spherical symmetry. After taking this into account, together with the specific requirements of moving ions, such as viscosity and electrophoretic effects, Onsager was able to derive a theoretical expression to account for the empirical relation known as Kohlrausch's Law, for the molar conductivity, Λ_m.

$$\Lambda_m = \Lambda_m^0 - K\sqrt{c}$$

Λ_m^0 is known as the limiting molar conductivity, K is an empirical constant and c is the electrolyte concentration. Limiting here means "at the limit of the infinite dilution"). Onsager's expression is

$$\Lambda_m = \Lambda_m^0 - (A + B\Lambda_m^0)\sqrt{c}$$

where A and B are constants that depend only on known quantities such as temperature, the charges on the ions and the dielectric constant and viscosity of the solvent. This is known as the Debye-Hückel-Onsager equation. However, this equation only applies to very dilute solutions and has been largely superseded by other equations due to Fuoss and Onsager, 1932 and 1957 and later.

Paschen's Law

Paschen's Law is an equation that gives the breakdown voltage, that is the voltage necessary to start a discharge or electric arc, between two electrodes in a gas as a function of pressure and gap length. It is named after Friedrich Paschen who discovered it empirically in 1889.

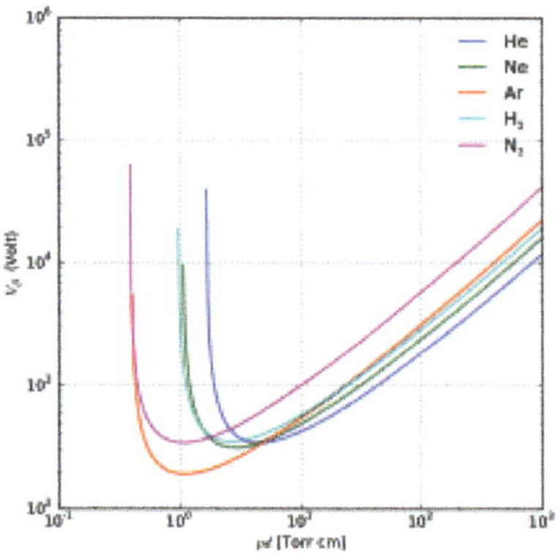

Paschen curves obtained for helium, neon, argon, hydrogen and nitrogen, using the expression for the breakdown voltage as a function of the parameters A,B that interpolate the first Townsend coefficient.

Paschen studied the breakdown voltage of various gases between parallel metal plates as the gas pressure and gap distance were varied. The voltage necessary to arc across the gap decreased as the pressure was reduced and then increased gradually, exceeding its original value. He also found that at normal pressure, the voltage needed to cause an arc reduced as the gap size was reduced but only to a point. As the gap was reduced further, the voltage required to cause an arc began to rise and again exceeded its original value. For a given gas, the voltage is a function only of the product of the pressure and gap length. The curve he found of voltage versus the pressure-gap length product *(right)* is called Paschen's curve. He found an equation that fit these curves, which is now called Paschen's law.

At higher pressures and gap lengths, the breakdown voltage is approximately *proportional* to the product of pressure and gap length, and the term Paschen's law is sometimes used to refer to this simpler relation. However this is only roughly true, over a limited range of the curve.

Paschen Curve

Early vacuum experimenters found a rather surprising behavior. An arc would sometimes take place in a long irregular path rather than at the minimum distance between the electrodes. For example, in air, at a pressure of 10^{-3} atmospheres, the distance for minimum breakdown voltage is about 7.5 μm. The voltage required to arc this distance is 327 V which is insufficient to ignite the arcs for gaps that are either wider or narrower. For a 3.5 μm gap, the required voltage is 533 V, nearly twice as much. If 500 V were applied, it would not be sufficient to arc at the 2.85 μm distance, but would arc at a 7.5 μm distance.

It was found that breakdown voltage was described by the equation:

$$V_B = \frac{Bpd}{\ln(Apd) - ln[ln(1+\frac{1}{\gamma_{se}})]}$$

Where V is the breakdown voltage in Volts, p is the pressure in Pascals, is the gap distance in meters, γ_{se} is the secondary electron emission coefficient at the cathode, A is the saturation ionization in the gas at a particular cE / p (electric field/pressure), and B is related to the excitation and ionization energies. The constants A and B are determined experimentally and found to be roughly constant over a restricted range of E / p for any given gas. For example, air with an E / p in the range of 450 to 7500 V·(kPa·cm)$^{-1}$, A = 112.50 (kPa·cm)$^{-1}$ and B = 2737.50 V·(kPa·cm)$^{-1}$. The graph of this equation is the Paschen curve. By differentiating it with respect to pd and setting the derivative to zero, the minimum voltage can be found. This yields

$$pd = e^{1-b}$$

and predicts the occurrence of a minimum breakdown voltage for $pd = 7.5\times10^{-6}$ m·atm. This is 327 V in air at standard atmospheric pressure at a distance of 7.5 µm. The composition of the gas determines both the minimum arc voltage and the distance at which it occurs. For argon, the minimum arc voltage is 137 V at a larger 12 µm. For sulfur dioxide, the minimum arc voltage is 457 V at only 4.4 µm.

For air at STP, the voltage needed to arc a 1-meter gap is about 3.4 MV. The intensity of the electric field for this gap is therefore 3.4 MV/m. The electric field needed to arc across the minimum voltage gap is much greater than what is necessary to arc a gap of one meter. For a 7.5 µm gap the arc voltage is 327 V which is 43 MV/m. This is about 13 times greater than the field strength for the 1 meter gap. The phenomenon is well verified experimentally and is referred to as the Paschen minimum. The equation loses accuracy for gaps under about 10 µm in air at one atmosphere and incorrectly predicts an infinite arc voltage at a gap of about 2.7 micrometers. Breakdown voltage can also differ from the Paschen curve prediction for very small electrode gaps when field emission from the cathode surface becomes important.

Physical Mechanism

The mean free path of a molecule in a gas is the average distance between its collision with other molecules. This is inversely proportional to the pressure of the gas. In air the mean free path of molecules is about 96 nm. Since electrons are much faster, their average distance between colliding with molecules is about 5.6 times longer or about 0.5 µm. This is a substantial fraction of the 7.5 µm spacing between the electrodes for minimum arc voltage. If the electron is in an electric field of 43 MV/m, it will be accelerated and acquire 21.5 electron volts of energy in 0.5 µm of travel in the direction of the field. The first ionization energy needed to dislodge an electron from nitrogen is about 15 eV. The accelerated electron will acquire more than enough energy to ionize a nitrogen atom. This liberated electron will in turn be accelerated which will lead to another collision. A chain reaction then leads to avalanche breakdown and an arc takes place from the cascade of released electrons.

More collisions will take place in the electron path between the electrodes in a higher pressure gas. When the pressure-gap product pd is high, an electron will collide with many different gas molecules as it travels from the cathode to the anode. Each of the collisions randomizes the electron direction, so the electron is not always being accelerated by the electric field—sometimes it travels back towards the cathode and is decelerated by the field.

Collisions reduce the electron's energy and make it more difficult for it to ionize a molecule. Energy losses from a greater number of collisions require larger voltages for the electrons to accumulate sufficient energy to ionize many gas molecules, which is required to produce an avalanche breakdown.

On the left side of the Paschen minimum, the pd product is small. The electron mean free path can become long compared to the gap between the electrodes. In this case, the electrons might gain lots of energy, but have fewer ionizing collisions. A greater voltage is therefore required to assure ionization of enough gas molecules to start an avalanche.

Derivation

Basics

To calculate the breakthrough voltage a homogeneous electrical field is assumed. This is the case in a parallel plate capacitor setup. The electrodes may have the distance d. The cathode is located at the point $x = 0$.

To get impact ionization the electron energy E_I must become greater than the ionization energy of the gas atoms between the plates. Per length of path x a number of α ionizations will occur. α is known as the first Townsend coefficient as it was introduced by Townsend in, section 17. The increase of the electron current Γ_e can be described for the assumed setup as

$$\Gamma_e(x = d) = \Gamma_e(x = 0)\mathrm{e}^{\alpha d} \qquad (1)$$

(So the number of free electrons at the anode is equal to the number of free electrons at the cathode that were multiplied by impact ionization. The larger d and/or α the more free electrons are created.)

The number of created electrons is

$$\Gamma_e(d) - \Gamma_e(0) = \Gamma_e(0)\left(\mathrm{e}^{\alpha d} - 1\right) \qquad (2)$$

Neglecting possible multiple ionizations of the same atom, the number of created ions is the same as the number of created electrons:

$$\Gamma_i(0) - \Gamma_i(d) = \Gamma_e(0)\left(\mathrm{e}^{\alpha d} - 1\right) \qquad (3)$$

Γ_i is the ion current. To keep the discharge going on, free electrons must be created at the cathode surface. This is possible because the ions hitting the cathode release secondary electrons at the impact. (For very large applied voltages also field electron emission can occur.) Without field emission, we can write

$$\Gamma_e(0) = \gamma\Gamma_i(0) \qquad (4)$$

where γ is the mean number of generated secondary electrons per ion. This is also known as the second Townsend coefficient. Assuming that $\Gamma_i(d) = 0$ one gets the relation between the Townsend coefficients by putting (4) into (3) and transforming:

$$\alpha d = \ln\left(1 + \frac{1}{\gamma}\right) \qquad (5)$$

Impact Ionization

What is the amount of ? The number of ionization depends upon the probability that an electron hits an ion. This probability P is the relation of the cross-sectional area of a collision between electron and ion σ in relation to the overall area A that is available for the electron to fly through:

$$P = \frac{N\sigma}{A} = \frac{x}{\lambda} \qquad (6)$$

As expressed by the second part of the equation, it is also possible to express the probability as relation of the path traveled by the electron x to the mean free path λ (distance at which another collision occurs).

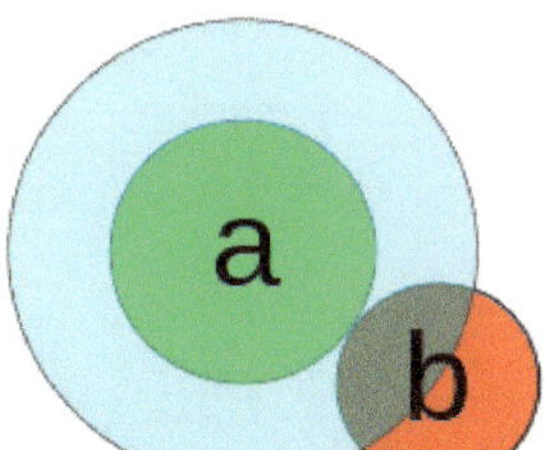

Visualization of the cross-section σ : If the center of particle b penetrates the blue circle, a collision occurs with particle a. So the area of the circle is the cross-section and its radius r is the sum of the radii of the particles.

N is the number of molecules which electrons can hit. It can be calculated using the equation of state of the ideal gas

$$pV = Nk_BT \qquad (7)$$

(p : pressure, V : volume, k_B : Boltzmann constant, T : temperature)

The adjoining sketch illustrates that $\sigma = \pi(r_a + r_b)^2$. As the radius of an electron can be neglected compared to the radius of an ion r_I it simplifies to $\sigma = \pi r_I^2$. Using this relation, putting (7) into (6) and transforming to λ one gets

$$\lambda = \frac{k_B T}{p \pi r_I^2} = \frac{1}{L \cdot p} \qquad (8)$$

where the factor L was only introduced for a better overview.

The alteration of the current of not yet collided electrons at every point in the path x can be expressed as

$$\mathrm{d}\Gamma_e(x) = -\Gamma_e(x) \frac{\mathrm{d}x}{\lambda_e} \qquad (9)$$

This differential equation can easily be solved:

$$\Gamma_e(x) = \Gamma_e(0) \exp\left(-\frac{x}{\lambda_e}\right) \qquad (10)$$

The probability that $\lambda > x$ (that there was not yet a collision at the point x) is

$$P(\lambda > x) = \frac{\Gamma_e(x)}{\Gamma_e(0)} = \exp\left(-\frac{x}{\lambda_e}\right) \qquad (11)$$

According to its definition α is the number of ionizations per length of path and thus the relation of the probability that there was no collision in the mean free path of the ions, and the mean free path of the electrons:

$$\alpha = \frac{P(\lambda > \lambda_I)}{\lambda_e} = \frac{1}{\lambda_e} \exp\left(0 \frac{\lambda_I}{\lambda_e}\right) = \frac{1}{\lambda_e} \exp\left(0 \frac{E_I}{E_e}\right) \qquad (12)$$

It was hereby considered that the energy E that a charged particle can get between a collision depends on the electric field strength $\mathcal{E}$ and the charge Q:

$$E = \lambda Q \mathcal{E} \qquad (13)$$

Breakdown Voltage

For the parallel-plate capacitor we have $\mathcal{E} = \frac{U}{d}$, where U is the applied voltage. As a single ionization was assumed Q is the elementary charge e. We can now put (13) and (8) into (12) and get

$$\alpha = L \cdot p \exp\left(0 \frac{L \cdot p \cdot d \cdot E_I}{eU}\right) \tag{14}$$

Putting this into (5) and transforming to U we get the Paschen law for the breakdown voltage $U_{\text{breakdown}}$ that was first investigated by Paschen in and whose formula was first derived by Townsend in, section 227:

$$U_{\text{breakdown}} = \frac{L \cdot p \cdot d \cdot E_I}{e\left(\ln(L \cdot p \cdot d) - \ln\left(\ln\left(1+\gamma^{-1}\right)\right)\right)} \tag{15}$$

$$\text{with } L = \frac{\pi r_I^2}{k_B T}$$

Plasma Ignition

Plasma ignition in definition of Townsend (Townsend discharge) is a self-sustaining discharge, independent of an external source of free electrons. This means that electrons from the cathode can reach the anode in the distance d and ionize at least one atom on its way. So according to the definition of α this relation must be fulfilled:

$$\alpha d \geq 1 \qquad (16) \alpha d \geq 1 \qquad (16)$$

If $\alpha d = 1$ is used instead of (5) one gets for the breakdown voltage

$$U_{\text{breakdown Townsend}} = \frac{L \cdot p \cdot d \cdot E_I}{e \cdot \ln(L \cdot p \cdot d)} = \frac{d \cdot E_I}{e \cdot \lambda_e \ln\left(\frac{d}{\lambda_e}\right)} \tag{17}$$

Conclusions / Validity

Paschen's law requires that

- there are already free electrons at the cathode ($\Gamma_e(x=0) \neq 0$) which can be accelerated to trigger impact ionization. Such so-called *seed electrons* can be created by ionization by cosmic x-ray background.
- the creation of further free electrons is only achieved by impact ionization. Thus Paschen's law is not valid if there are external electron sources. This can for example be a light source creating secondary electrons via the photoelectric effect. This has to be considered in experiments.
- each ionized atom leads to only one free electron. But multiple ionizations occur always in practice.

- free electrons at the cathode surface are created by the impacting ions. The problem is that the number of thereby created electrons strongly depends on the material of the cathode, its surface (roughness, impurities) and the environmental conditions (temperature, humidity etc.). The experimental, reproducible determination of the factor is therefore nearly impossible.
- the electrical field is homogeneous.

Effects with Different Gases

Different gases will have different mean free paths for molecules and electrons. This is because different molecules have different diameters. Noble gases like helium and argon are monatomic and tend to have smaller diameters. This gives them a greater mean free path length.

Ionization potentials differ between molecules as well as the speed that they recapture electrons after they have been knocked out of orbit. All three effects change the number of collisions needed to cause an exponential growth in free electrons. These free electrons are necessary to cause an arc.

Faraday's Laws of Electrolysis

Faraday's laws of electrolysis are quantitative relationships based on the electrochemical researches published by Michael Faraday in 1834. In the same years, in Italy, Carlo Matteucci discovered the laws of electrolysis by a method totally independent Faraday's methods. The laws of electrolysis can also be called Faraday-Matteucci's laws.

Michael Faraday.

Mathematical Form

Faraday's laws can be summarized by

$$m = \left(\frac{Q}{F}\right)\left(\frac{M}{z}\right)$$

where:

- m is the mass of the substance liberated at an electrode in grams
- Q is the total electric charge passed through the substance in coulombs
- $F = 96485\ \mathrm{C\ mol^{-1}}$ is the Faraday constant
- M is the molar mass of the substance in grams per mol
- z is the valency number of ions of the substance (electrons transferred per ion).

Note that M/z is the same as the equivalent weight of the substance altered.

For Faraday's first law, M, F, and z are constants, so that the larger the value of Q the larger m will be.

For Faraday's second law, Q, F, and z are constants, so that the larger the value of M/z (equivalent weight) the larger m will be.

In the simple case of constant-current electrolysis, $Q = It$ leading to

$$m = \left(\frac{It}{F}\right)\left(\frac{M}{z}\right)$$

and then to

$$n = \left(\frac{It}{F}\right)\left(\frac{1}{z}\right)$$

where:

- n is the amount of substance ("number of moles") liberated: $n = m/M$
- t is the total time the constant current was applied.

In the more complicated case of a variable electric current, the total charge Q is the electric current $I(\tau)$ integrated over time τ:

$$Q = \int_0^t I(\tau)\, d\tau$$

Here t is the *total* electrolysis time.

References

- Lieberman, Michael A.; Lichtenberg, Allan J. (2005). Principles of plasma discharges and materials processing (2nd ed.). Hoboken, N.J.: Wiley-Interscience. 546. ISBN 978-0471005773. OCLC 59760348.
- Wadhwa, C.L. (2007). High Voltage Engineering (2nd ed.). New Age International. pp. 10–12. ISBN 8122418597.
- Graf, Rudolf F. (1999). Modern Dictionary of Electronics (7th ed.). Newnes. p. 542. ISBN 0750698667.

7

Major Aspects of Electrochemistry

Electrochemistry is best understood in confluence with the major topics listed in the following chapter. Some of these topics are electrode potential, absolute electrode potential, standard electrode potential, molar conductivity and electrochemical potential. This text helps the reader in developing an in-depth understanding of the major aspects of electrochemistry.

Electrode Potential

Electrode potential, *E*, in electrochemistry, according to an IUPAC definition, is the electromotive force of a cell built of two electrodes:

- on the left-hand side is the standard hydrogen electrode, and
- on the right-hand side is the electrode the potential of which is being defined.

By convention:

$$E_{\text{Cell}} = E_{\text{Cathode}} - E_{\text{Anode}}$$

From the above, for the cell with the standard hydrogen electrode (potential of 0 by convention), one obtains:

$$E_{\text{Cell}} = E_{\text{Right}} - 0 = E_{\text{Electrode}}$$

The left-right convention is consistent with the international agreement that redox potentials be given for reactions written in the form of reduction half-reactions.

Electrode potential is measured in volts (V).

Origin and Interpretation

Electrode potential appears at the interface between an electrode and electrolyte due to the transfer of charged species across the interface, specific adsorption of ions at the interface, and specific adsorption/orientation of polar molecules, including those of the solvent.

Electrode potential is the electric potential on an electrode component. In a cell, there will be an electrode potential for the cathode, and an electrode potential for the anode. The difference between the two electrode potentials equals the cell potential.

$$E_{\text{Cell}} = E_{\text{Cathode}} - E_{\text{Anode}}$$

The measured electrode potential may be either that at equilibrium on the working electrode ("reversible potential"), or a potential with a non-zero net reaction on the working electrode but zero net current ("corrosion potential", "mixed potential"), or a potential with a non-zero net current on the working electrode (like in galvanic corrosion or voltammetry). Reversible potentials can be sometimes converted to the standard electrode potential for a given electroactive species by extrapolation of the measured values to the standard state.

The value of the electrode potential under non-equilibrium depends on the nature and composition of the contacting phases, and on the kinetics of electrode reactions at the interface.

An operational assumption for determinations of the electrode potentials with the standard hydrogen electrode involves this reference electrode with hydrogen ion in an ideal solution having is "zero potential at all temperatures" equivalently to standard enthalpy of formation of hydrogen ion is also "zero at all temperatures".

Measurement

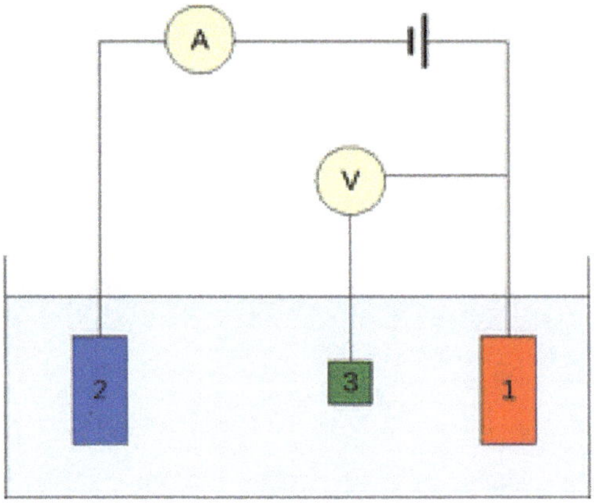

Three-electrode setup for measurement of electrode potential.

The measurement is generally conducted using a three-electrode setup:

1. Working electrode
2. Counter electrode
3. Reference electrode (standard hydrogen electrode or an equivalent)

In case of non-zero net current on the electrode, it is essential to minimize the ohmic IR-drop in the electrolyte, e.g., by positioning the reference electrode near the surface of the working electrode, or by using a supporting electrolyte of sufficiently high conductivity. The potential measurements are performed with the positive terminal of the electrometer connected to working electrode and the negative terminal to the reference electrode.

Sign Conventions

Historically, two conventions for sign for the electrode potential have formed:

1. Convention "Nernst-Lewis-Latimer" (sometimes referred to as "American");
2. Convention "Gibbs-Ostwald-Stockholm" (sometimes referred to as "European").

In 1953 in Stockholm IUPAC recognized that either of the conventions is permissible; however, it unanimously recommended that only the magnitude expressed according to the convention (2) be called "the electrode potential". To avoid possible ambiguities, the electrode potential thus defined can also be referred to as *Gibbs-Stockholm electrode potential.*

In both conventions, the EMF of the half-cell is referred to the standard hydrogen electrode as the zero point.

According to the Nernst-Lewis-Latimer convention (1), the sign of the potential is determined by the change of the sign of the standard Gibbs free energy of the half-cell reaction. This reaction transfers nF of the positive electric charge across the phase boundary from the left to the right of the half-cell, as shown in the half-cell diagram. The relation is written as:

$$\Delta G = -nFE$$

where: ΔG - the standard Gibbs energy for the half-cell reaction, n - number of electrons transferred across the across the phase boundary in the above reaction, F - the Faraday constant, E - the half-cell potential according to the Nernst-Lewis-Latimer convention.

Therefore, according to the convention (1), the value of EMF is bivariant (with two variants), i.e., reversing the electrode reaction, and reversing the corresponding cell diagram, will cause the change of sign of ΔG and E, even though the polarity of the cell measured with a direct-current voltmeter does not necessarily reverse. This convention has been historically (sometimes still is) used to express anodic oxidation potentials or cathodic reduction potentials of half-cells. The reduction potentials expressed according to (1) have the same sign as those expressed according to the convention (2).

According to the Gibbs-Stockholm convention (2), the sign of the electrode potential is determined by the polarity of the cell terminal measured against the standard hydrogen electrode using an instrument for DC-potential measurements, when this circuit is closed and balanced (i.e., the current on the measurement circuit is nil). In convention (2), the value of the measured EMF is invariant of the direction of the infinitesimally small changes of the electric current that may be allowed on the measured electrode and render it either cathodic or anodic. Therefore, the value of E does not change with the direction of the (very small) current, even though ΔG of the corresponding reaction is bivariant (can be of either positive or negative sign).

Potential Difference of a Cell Assembled of Two Electrodes

Potential of a cell assembled of two electrodes can be determined from the two individual electrode potentials using:

$$\Delta V_{\text{cell}} = E_{\text{red,cathode}} - E_{\text{red,anode}}$$

or, equivalently,

$$\Delta V_{\text{cell}} = E_{\text{red,cathode}} + E_{\text{oxy,anode}}$$

This follows from the IUPAC definition of the electric potential difference of a galvanic cell, according to which the electric potential difference of a cell is the difference of the potentials of the electrodes on the right and the left of the galvanic cell. When ΔV_{cell} is positive, then positive electrical charge flows through the cell from the left electrode (anode) to the right electrode (cathode).

Absolute Electrode Potential

Absolute electrode potential, in electrochemistry, according to an IUPAC definition, is the electrode potential of a metal measured with respect to a universal reference system (without any additional metal–solution interface).

Definition

According to a more specific definition presented by Trasatti, the absolute electrode potential is the difference in electronic energy between a point inside the metal (Fermi level) of an electrode and a point outside the electrolyte in which the electrode is submerged (an electron at rest in vacuum).

This potential is difficult to determine accurately. For this reason, standard hydrogen electrode is typically used for reference potential. The absolute potential of the SHE is 4.44 ± 0.02 V at 25 °C. Therefore, for any electrode at 25 °C:

$$E^{M}_{(\text{abs})} = E^{M}_{(\text{SHE})} + (4.44 \pm 0.02)\ \text{V}$$

where:

E is electrode potential

V is volt

M denotes the electrode made of metal M

(abs) denotes the absolute potential

(SHE) denotes the electrode potential relative to the standard hydrogen electrode.

A different definition for the absolute electrode potential (also known as absolute half-cell potential and single electrode potential) has also been discussed in the literature. In this approach, one first defines an isothermal absolute single-electrode process (or absolute half-cell process.) For example, in the case of a generic metal being oxidized to form a solution-phase ion, the process would be

$$M_{(metal)} \rightarrow M^{+}_{(solution)} + e-_{(gas)}$$

For the hydrogen electrode, the absolute half-cell process would be

$$1/2H_{2\ (gas)} \rightarrow H^{+}_{(solution)} + e-_{(gas)}$$

Other types of absolute electrode reactions would be defined analogously.

In this approach, all three species taking part in the reaction, including the electron, must be placed in thermodynamically well-defined states. All species, including the electron, are at the same temperature, and appropriate standard states for all species, including the electron, must be fully defined. The absolute electrode potential is then defined as the Gibbs free energy for the absolute electrode process. To express this in volts one divides the Gibb's free energy by the negative of Faraday's constant.

Rockwood's approach to absolute-electrode thermodynamics is easily expendable to other thermodynamic functions. For example, the absolute half-cell entropy has been defined as the entropy of the absolute half-cell process defined above. An alternative definition of the absolute half-cell entropy has recently been published by Fang et al. who define it as the entropy of the following reaction (using the hydrogen electrode as an example):

$$1/2H_{2\ (gas)} \rightarrow H^{+}_{(solution)} + e-_{(metal)}$$

This approach differs from the approach described by Rockwood in the treatment of the electron, i.e. whether it is placed in the gas phase or in the metal.

Determination

The basis for determination of the absolute electrode potential under the Trasatti definition is given by the equation:

$$E^{M}\,(\text{abs}) = \phi^{M} + \Delta_{S}^{M}\psi$$

where:

E^M(abs) is the absolute potential of the electrode made of metal M

ϕ^M is the electron work function of metal M

$\Delta_S^M \psi$ is the contact (Volta) potential difference at the metal(M)–solution(S) interface.

For practical purposes, the value of the absolute electrode potential of the standard hydrogen electrode is best determined with the utility of data for an ideally-polarizable mercury (Hg) electrode:

$$E^{\ominus}(H^+/H_2)(abs) = \phi^{Hg} + \Delta_S^{Hg}\psi^{\ominus}_{\sigma=0} - E^{Hg}_{\sigma=0}(SHE)$$

where:

$E^{\ominus}(H^+/H_2)$(abs) is the absolute standard potential of the hydrogen electrode

$\sigma = 0$ denotes the condition of the point of zero charge at the interface.

The types of physical measurements required under the Rockwood definition are similar to those required under the Trasatti definition, but they are used in a different way, e.g. in Rockwood's approach they are used to calculate the equilibrium vapor pressure of the electron gas. The numerical value for the absolute potential of the standard hydrogen electrode one would calculate under the Rockwood definition is sometimes fortuitously close to the value one would obtain under the Trasatti definition. This near-agreement in the numerical value depends on the choice of ambient temperature and standard states, and is the result of the near-cancellation of certain terms in the expressions. For example, if a standard state of one atmosphere ideal gas is chosen for the electron gas then the cancellation of terms occurs at a temperature of 296 K, and the two definitions give an equal numerical result. At 298.15 K a near-cancellation of terms would apply and the two approaches would produce nearly the same numerical values. However, there is no fundamental significance to this near agreement because it depends on arbitrary choices, such as temperature and definitions of standard states.

Standard Electrode Potential

In electrochemistry, the standard electrode potential, abbreviated *E*° or E⦵ (with a superscript plimsoll character, pronounced "standard" or "nought"), is the measure of individual potential of a reversible electrode at standard state, which is with solutes at an effective concentration of 1 mol dm^{-3}, and gases at a pressure of 1 atm. The reduction potential is an intensive property. The values are most often tabulated at 25 °C. The basis for an electrochemical cell such as the galvanic cell is always a redox reaction which can be broken down into two half-reactions: oxidation at anode (loss of electron) and reduction at cathode (gain of electron). Electricity is generated due to electric potential difference between two electrodes. This potential difference is created as a result of the difference between individual potentials of the two metal electrodes with respect to the electrolyte. (Reversible electrode is an electrode that owes its potential to changes of a reversible nature, in contrast to electrodes used in electroplating which are destroyed during their use.)

Although the overall potential of a cell can be measured, there is no simple way to accurately measure the electrode/electrolyte potentials in isolation. The electric potential also varies with temperature, concentration and pressure. Since the oxidation potential of a half-reaction is the negative of the reduction potential in a redox reaction, it is sufficient to calculate either one of the potentials. Therefore, standard electrode potential is commonly written as standard reduction potential.

Calculation of Standard Electrode Potentials

The electrode potential cannot be obtained empirically. The galvanic cell potential results from a *pair* of electrodes. Thus, only one empirical value is available in a pair of electrodes and it is not possible to determine the value for each electrode in the pair using the empirically obtained galvanic cell potential. A reference electrode, standard hydrogen electrode (SHE), for which the potential is *defined* or agreed upon by convention, needed to be established. In this case SHE is set to 0.00 V and any electrode, for which the electrode potential is not yet known, can be paired with SHE—to form a galvanic cell—and the galvanic cell potential gives the unknown electrode's potential. Using this process, any electrode with an unknown potential can be paired with either the SHE or another electrode for which the potential has already been derived and that unknown value can be established.

Since the electrode potentials are conventionally defined as reduction potentials, the sign of the potential for the metal electrode being oxidized must be reversed when calculating the overall cell potential. Note that the electrode potentials are independent of the number of electrons transferred —they are expressed in volts, which measure energy per electron transferred—and so the two electrode potentials can be simply combined to give the overall *cell* potential even if different numbers of electrons are involved in the two electrode reactions.

For practical measurements, the electrode in question is connected to the positive terminal of the electrometer, while SHE is connected to the negative terminal.

Standard Reduction Potential Table

The larger the value of the standard reduction potentials, the easier it is for the element to be reduced (accept electrons); in other words, they are better oxidizing agents. For example, F_2 has 2.87 V and Li^+ has −3.05 V. F reduces easily and is therefore a good oxidizing agent. In contrast, $Li_{(s)}$ would rather undergo oxidation (hence a good reducing agent). Thus Zn^{2+} whose standard reduction potential is −0.76 V can be oxidized by any other electrode whose standard reduction potential is greater than −0.76 V (e.g. H^+(0 V), Cu^{2+}(0.16 V), F_2(2.87 V)) and can be reduced by any electrode with standard reduction potential less than −0.76 V (e.g. H_2(−2.23 V), Na^+(−2.71 V), Li^+(−3.05 V)).

In a galvanic cell, where a spontaneous redox reaction drives the cell to produce an electric potential, Gibbs free energy ΔG^o must be negative, in accordance with the following equation:

$$\Delta G^o_{cell} = -nFE^o_{cell}$$

where n is number of moles of electrons per mole of products and F is the Faraday constant, ~96485 C/mol. As such, the following rules apply:

If $E^o_{cell} > 0$, then the process is spontaneous (galvanic cell)

If $E^o_{cell} < 0$, then the process is nonspontaneous (electrolytic cell)

Thus in order to have a spontaneous reaction ($\Delta G^o < 0$), E^o_{cell} must be positive, where:

$$E^o_{cell} = E^o_{cathode} - E^o_{anode}$$

where E^o_{anode} is the standard potential at the anode and $E^o_{cathode}$ is the standard potential at the cathode as given in the table of standard electrode potential.

Molar Conductivity

Molar conductivity is defined as the conductivity of an electrolyte solution divided by the molar concentration of the electrolyte, and so measures the efficiency with which a given electrolyte conducts electricity in solution. It is the conducting power of all the ions produced by dissolving one mole of an electrolyte in solution. Its units are siemens per meter per molarity, or siemens meter-squared per mole. The usual symbol is a capital lambda, Λ, or Λ_m. It can also be defined as follows: Molar conductivity of a solution at a given concentration is the conductance of the volume (V) of the solution containing one mole of electrolyte kept between two electrodes with area of cross section (A) and at a distance of unit length.

History

Friedrich Kohlrausch's researches in 1875-79 established that to a high accuracy in dilute solutions, molar conductivity is composed of individual contributions of ions. This is known as the *law of independent migration of ions*.

Description

From its definition, the molar conductivity is given by:

$$\Lambda_m = \frac{\kappa}{c}$$

where:

- κ is the measured conductivity
- c is the electrolyte concentration.

Two cases should be distinguished: strong electrolytes and weak electrolytes.

For strong electrolytes, such as salts, strong acids and strong bases, the molar conductivity depends only *weakly* on concentration. Based on experimental data Friedrich Kohlrausch (around the year 1900) proposed the non-linear law for strong electrolytes:

$$\Lambda_m = \Lambda_m^{\circ} - K\sqrt{c}$$

where

- Λ_m° is the molar conductivity at infinite dilution (or *limiting molar conductivity*), which can be determined by extrapolation of Λ_m as a function of $\sqrt{c}$
- K is the Kohlrausch coefficient, which depends mainly on the stoichiometry of the specific salt in solution.

This law is valid for low electrolyte concentrations only; it fits into the Debye-Hückel-Onsager equation :.

For weak electrolytes (i.e. incompletely dissociated electrolytes), however, the molar conductivity *strongly* depends on concentration: The more dilute a solution, the greater its *molar* conductivity, due to increased ionic dissociation. (This, for example, is the case of SDS-coated proteins in the stacking gel of an SDS-PAGE.)

Molar Ionic Conductivity

The limiting molar conductivity can be decomposed into contributions from the different ions (Kohlrausch's law of independent migration of ions):

$$\Lambda_m^{\circ} = \Sigma_i \nu_i \lambda_i$$

where:

- λ_i is the limiting molar ionic conductivity of ion *i*.

- ν_i is the number of ions i in the formula unit of the electrolyte (e.g. 2 and 1 for Na^+ and SO_4^{2-} in Na_2SO_4)

The molar ionic conductivity of each ionic species is proportional to its electrical mobility (), or drift velocity per unit electric field, according to the equation , where z is the ionic charge and F is the Faraday constant.

The limiting molar conductivity of a weak electrolyte cannot be determined reliably by extrapolation. Instead it can be expressed as a sum of ionic contributions which can be evaluated from the limiting molar conductivities of strong electrolytes containing the same ions. For aqueous acetic acid as an example,

$$\Lambda_m^\circ(CH_3COOH) = \lambda(CH_3COO^-) + \lambda(H^+) = \Lambda_m^\circ(CH_3COONa) + \Lambda_m^\circ(HCl) - \Lambda_m^\circ(NaCl)$$

Values for each ion may be determined using measured ion transport numbers. For the cation

- $\lambda^+ = t_+ \cdot \frac{\Lambda_0}{\nu^+}$ and for the anion $\lambda^- = t_- \cdot \frac{\Lambda_0}{\nu^-}$

Most monovalent ions in water have limiting molar ionic conductivities in the range of 40–80 S cm² mol⁻¹, for example cations Na^+ 50.1, K^+ 73.5 and Ag^+ 61.9, and anions F^- 55.4, Br^- 78.1 and CH_3COO^- 40.9. Exceptionally high values are found for H^+ (349.8) and OH^- (198.6), which are explained by the Grotthuss proton-hopping mechanism for the movement of these ions. The H^+ also has a larger conductivity than other ions in alcohols which have a hydroxyl group, but behaves more normally in other solvents including liquid ammonia and nitrobenzene.

For multivalent ions, it is usual to consider the conductivity divided by the equivalent ion concentration in terms of equivalents per litre, where 1 equivalent is the quantity of ions which have the same amount of electric charge as 1 mol of a monovalent ion: 1/2 mol Ca^{2+}, 1/2 mol SO_4^{2-}, 1/3 mol Al^{3+}, 1/4 mol $Fe(CN)_6^{4-}$, etc. This quotient can be called the *equivalent conductivity*, although IUPAC has recommended that use of this term be discontinued and the term molar conductivity be used for the values of conductivity divided by equivalent concentration. If this convention is used, then the values are in the same range as monovalent ions, e.g. 59.5 S cm² mol⁻¹ for 1/2 Ca^{2+} and 80.0 for 1/2 SO_4^{2-}.

From the ionic molar conductivities of cations and anions, ionic radii can be calculated using the concept of Stokes radius. The values obtained for an ionic radius in solution calculated this way can be quite different from the ionic radius for the same ion in crystals, due to the effect of hydration in solution.

Applications

Ostwald's law of dilution, which gives the dissociation constant of a weak electrolyte as a function of concentration, can be written in terms of molar conductivity. Thus, the pKa values of acids can be calculated by measuring the molar conductivity and extrapolating to zero concentration. Namely, $pK_a = p(K/(1 \text{ mol dm}^{-3}))$ at the zero-concentration limit, where K is the dissociation constant from Ostwald's law.

Electrochemical Potential

In electrochemistry, the electrochemical potential, μ, sometimes abbreviated to ECP, is a thermodynamic measure of chemical potential that does not omit the energy contribution of electrostatics. Electrochemical potential is expressed in the unit of J/mol.

Introduction

Each chemical species (for example, "water molecules", "sodium ions", "electrons", etc.) has an electrochemical potential (a quantity with units of energy) at any given point in space, which represents how easy or difficult it is to add more of that species to that location. If possible, a species will move from areas with higher electrochemical potential to areas with lower electrochemical potential; in equilibrium, the electrochemical potential will be constant everywhere for each species (it may have a different value for different species). For example, if a glass of water has sodium ions (Na^+) dissolved uniformly in it, and an electric field is applied across the water, then the sodium ions will tend to get pulled by the electric field towards one side. We say the ions have electric potential energy, and are moving to lower their potential energy. Likewise, if a glass of water has a lot of dissolved sugar on one side and none on the other side, each sugar molecule will randomly diffuse around the water, until there is equal concentration of sugar everywhere. We say that the sugar molecules have a "chemical potential," which is higher in the high-concentration areas, and the molecules move to lower their chemical potential. These two examples show that an electrical potential and a chemical potential can both give the same result: A redistribution of the chemical species. Therefore, it makes sense to combine them into a single "potential", the *electrochemical potential*, which can directly give the *net* redistribution taking *both* into account.

It is (in principle) easy to measure whether or not two regions (for example, two glasses of water) have the same electrochemical potential for a certain chemical species (for example, a solute molecule): Allow the species to freely move back and forth between the two regions (for example, connect them with a semi-permeable membrane that lets only that species through). If the chemical potential is the same in the two regions, the species will occasionally move back and forth between the two regions, but on average there is just as much movement in one direction as the other, and there is zero net migration (this is called "diffusive equilibrium"). If the chemical potentials of the two regions are different, more molecules will move to the lower chemical potential than the other direction.

Moreover, when there is *not* diffusive equilibrium, i.e., when there is a tendency for molecules to diffuse from one region to another, then there is a certain free energy released by each net-diffusing molecule. This energy, which can sometimes be harnessed (a simple example is a concentration cell), and the free-energy per mole is exactly equal to the electrochemical potential difference between the two regions.

Conflicting Terminologies

It is common in electrochemistry and solid-state physics to discuss both the chemical potential and the electrochemical potential of the electrons. However, in the two fields, the definitions of these two terms are sometimes swapped. In electrochemistry, the *electrochemical potential* of electrons (or any other species) is the total potential, including both the (internal, nonelectrical) chemical potential and the electric potential, and is by definition constant across a device in equilibrium, whereas the *chemical potential* of electrons is equal to the electrochemical potential minus the local electric potential energy per electron. In solid-state physics, the definitions are normally compatible with this, but occasionally the definitions are swapped.

Definition and Usage

In generic terms, electrochemical potential is the mechanical work done in bringing 1 mole of an ion from a standard state to a specified concentration and electrical potential. According to the IUPAC definition, it is the partial molar Gibbs energy of the substance at the specified electric potential, where the substance is in a specified phase. Electrochemical potential can be expressed as

$$\bar{\mu}_i = \mu_i + z_i F \Phi,$$

where:

- μ_i is the electrochemical potential of species i, in J/mol
- μ_i is the chemical potential of the species i, in J/mol
- z_i is the valency (charge) of the ion i, a dimensionless integer
- F is the Faraday constant, in C/mol
- Φ is the local electrostatic potential, in V.

In the special case of an uncharged atom, $z_i = 0$ and so $\mu_i = \mu_i$.

Electrochemical potential is important in biological processes that involve molecular diffusion across membranes, in electroanalytical chemistry, and industrial applications such as batteries and fuel cells. It represents one of the many interchangeable forms of potential energy through which energy may be conserved.

In cell membranes, the electrochemical potential is the sum of the chemical potential and the membrane potential.

Incorrect Usage

The term *electrochemical potential* is sometimes used to mean an electrode potential (either of a corroding electrode, an electrode with a non-zero net reaction or current, or an electrode at equilibrium). In some contexts, the electrode potential of corroding metals is called "electrochemical corrosion potential", which is often abbreviated as ECP, and the word "corrosion" is sometimes omitted. This usage can lead to confusion. The two quantities have different meanings and different dimensions: the dimension of electrochemical potential is energy per mole while that of electrode potential is voltage (energy per charge).

Volta Potential

The Volta potential (also called Volta potential difference, contact potential difference, outer potential difference, $\Delta\psi$, or "delta psi") in electrochemistry, is the electrostatic potential difference between two metals (or one metal and one electrolyte) that are in contact and are in thermodynamic equilibrium. Specifically, it is the potential difference between a point close to the surface of the first metal, and a point close to the surface of the second metal (or electrolyte).

The Volta potential is named after Alessandro Volta.

Volta Potential between Two Metals

When two metals are electrically isolated from each other, an arbitrary potential difference may exist between them. However, when two different neutral metal surfaces are brought into electrical contact (even indirectly, say, through a long wire), electrons will flow from the metal with a higher Fermi level to the metal with the lower Fermi level until the Fermi levels in the both phases are equal. Once this has occurred, the metals are in thermodynamic equilibrium with each other (the actual number of electrons that passes between the two phases is usually small). Just because the Fermi levels are equal, however, does not mean that the electric potentials are equal. The electric potential outside each material is controlled by its work function, and so dissimilar metals can show an electric potential difference even at equilibrium.

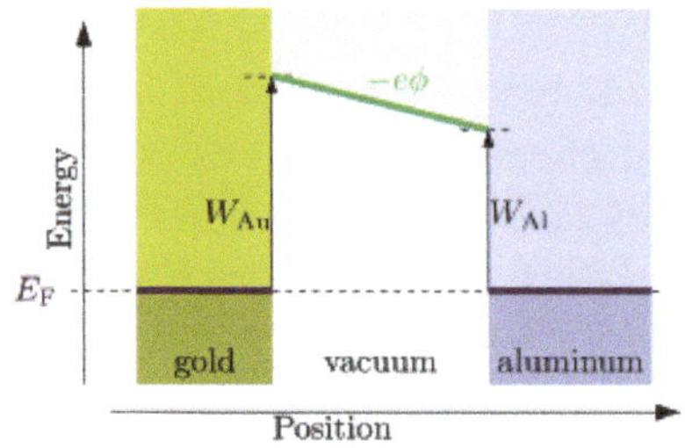

When the two metals depicted here are in thermodynamic equilibrium with each other as shown (equal Fermi levels), the vacuum electrostatic potential □ is not flat due to a difference in work function.

The Volta potential is *not* an intrinsic property of the two bulk metals under consideration, but rather is determined by work function differences between the metals' surfaces. Just like the work function, the Volta potential depends sensitively on surface state, contamination, and so on.

The potential is given by the formula: $V_{AB} = \frac{RT}{F} ln \frac{e_A}{e_B}$,

where V_{AB} is the potential,

- R is the gas constant,
- T absolute temperature,
- e_A, e_B, electron number densities in metals A and B.

Measurement of Volta Potential (Kelvin Probe)

The Volta potential can be significant (of order 1 volt) but it cannot be measured directly by an ordinary voltmeter. A voltmeter does not measure vacuum electrostatic potentials, but instead the difference in Fermi level between the two materials, a difference that is exactly zero at equilibrium.

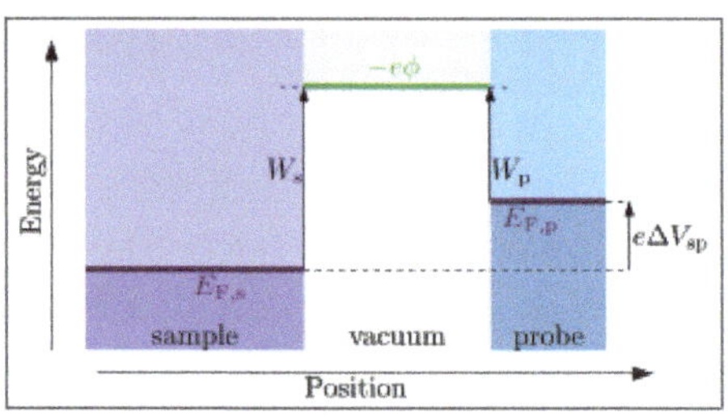

Kelvin probe energy diagram at flat vacuum configuration, used for measuring Volta potential between sample and probe.

The Volta potential however corresponds to a real electric field in the spaces between and around the two metal objects, a field generated by the accumulation of charges at their surfaces. The total charge over each objects' surface depends on the capacitance C between the two objects, by the relation $Q = C\Delta\psi$ where $\Delta\psi$ is the Volta potential. It follows therefore that the value of the potential can be measured by varying the capacitance between the materials by a known amount (e.g., by moving the objects further from each other), and measuring the displaced charge that flows through the wire that connects them.

The Volta potential difference between a metal and an electrolyte can be measured in a similar fashion. The Volta potential of a metal surface can be mapped on very small scales by use of a Kelvin probe force microscope. In this case the capacitance change is not known—instead, a compensating DC voltage is added to cancel the Volta potential so that no current is induced by the change in capacitance. This compensating voltage is the negative of the Volta potential.

References

- The best test preparation for the GRE Graduate Record Examination Chemistry Test. Published by the Research and Education Association, 2000, ISBN 0-87891-600-8. p. 149.

8

Corrosion: An Integrated Study

Corrosion is a process that occurs naturally, and helps in the conversion of metal to chemically stable form. It is the slow demolition of materials by chemicals. This section will provide an integrated understanding of corrosion.

Corrosion

Corrosion is a natural process, which converts a refined metal to a more stable form, such as its oxide, hydroxide, or sulfide. It is the gradual destruction of materials (usually metals) by chemical and/or electrochemical reaction with their environment. Corrosion engineering is the field dedicated to controlling and stopping corrosion.

Rust, the most familiar example of corrosion.

In the most common use of the word, this means electrochemical oxidation of metal in reaction with an oxidant such as oxygen or sulfur. Rusting, the formation of iron oxides, is a well-known example of electrochemical corrosion. This type of damage typically produces oxide(s) or salt(s) of the original metal, and results in a distinctive orange colouration. Corrosion can also occur in materials other than metals, such as ceramics or polymers, although in this context, the term "degradation" is more common. Corrosion degrades the useful properties of materials and structures including strength, appearance and permeability to liquids and gases.

Many structural alloys corrode merely from exposure to moisture in air, but the process can be strongly affected by exposure to certain substances. Corrosion can be concentrated locally to form a pit or crack, or it can extend across a wide area more or less uniformly corroding the surface. Because corrosion is a diffusion-controlled process, it occurs on exposed surfaces. As a result, methods to reduce the activity of the exposed surface, such as passivation and chromate conversion, can increase a material's corrosion resistance. However, some corrosion mechanisms are less visible and less predictable.

Volcanic gases have accelerated the corrosion of this abandoned mining machinery.

Galvanic Corrosion

Galvanic corrosion occurs when two different metals have physical or electrical contact with each other and are immersed in a common electrolyte, or when the same metal is exposed to electrolyte with different concentrations. In a galvanic couple, the more active metal (the anode) corrodes at an accelerated rate and the more noble metal (the cathode) corrodes at a slower rate. When immersed separately, each metal corrodes at its own rate. What type of metal(s) to use is readily determined by following the galvanic series. For example, zinc is often used as a sacrificial anode for steel structures. Galvanic corrosion is of major interest to the marine industry and also anywhere water (containing salts) contacts pipes or metal structures.

Galvanic corrosion of aluminium. A 5-mm-thick aluminium alloy plate is physically (and hence, electrically) connected to a 10-mm-thick mild steel structural support. Galvanic corrosion occurred on the aluminium plate along the joint with the steel. Perforation of aluminium plate occurred within 2 years.

Factors such as relative size of anode, types of metal, and operating conditions (temperature, humidity, salinity, etc.) affect galvanic corrosion. The surface area ratio of the anode and cathode directly affects the corrosion rates of the materials. Galvanic corrosion is often prevented by the use of sacrificial anodes.

Galvanic Series

In any given environment (one standard medium is aerated, room-temperature seawater), one metal will be either more *noble* or more *active* than others, based on how strongly its ions are bound to the surface. Two metals in electrical contact share the same electrons, so that the "tug-of-war" at each surface is analogous to competition for free electrons between the two materials. Using the electrolyte as a host for the flow of ions in the same direction, the noble metal will take electrons from the active one. The resulting mass flow or electric current can be measured to establish a hierarchy of materials in the medium of interest. This hierarchy is called a *galvanic series* and is useful in predicting and understanding corrosion.

Corrosion Removal

Often it is possible to chemically remove the products of corrosion. For example, phosphoric acid in the form of naval jelly is often applied to ferrous tools or surfaces to remove rust. Corrosion removal should not be confused with electropolishing, which removes some layers of the underlying metal to make a smooth surface. For example, phosphoric acid may also be used to electropolish copper but it does this by removing copper, not the products of copper corrosion.

Resistance to Corrosion

Some metals are more intrinsically resistant to corrosion than others. There are various ways of protecting metals from corrosion (oxidation) including painting, hot dip galvanizing, and combinations of these.

Intrinsic Chemistry

Gold nuggets do not naturally corrode, even on a geological time scale.

The materials most resistant to corrosion are those for which corrosion is thermodynamically unfavorable. Any corrosion products of gold or platinum tend to decompose spontaneously into pure metal, which is why these elements can be found in metallic form on Earth and have long been valued. More common "base" metals can only be protected by more temporary means.

Some metals have naturally slow reaction kinetics, even though their corrosion is thermodynamically favorable. These include such metals as zinc, magnesium, and cadmium. While corrosion of these metals is continuous and ongoing, it happens at an acceptably slow rate. An extreme example is graphite, which releases large amounts of energy upon oxidation, but has such slow kinetics that it is effectively immune to electrochemical corrosion under normal conditions.

Passivation

Passivation refers to the spontaneous formation of an ultrathin film of corrosion products, known as a passive film, on the metal's surface that act as a barrier to further oxidation. The chemical composition and microstructure of a passive film are different from the underlying metal. Typical passive film thickness on aluminium, stainless steels, and alloys is within 10 nanometers. The passive film is different from oxide layers that are formed upon heating and are in the micrometer thickness range – the passive film recovers if removed or damaged whereas the oxide layer does not. Passivation in natural environments such as air, water and soil at moderate pH is seen in such materials as aluminium, stainless steel, titanium, and silicon.

Passivation is primarily determined by metallurgical and environmental factors. The effect of pH is summarized using Pourbaix diagrams, but many other factors are influential. Some conditions that inhibit passivation include high pH for aluminium and zinc, low pH or the presence of chloride ions for stainless steel, high temperature for titanium (in which case the oxide dissolves into the metal, rather than the electrolyte) and fluoride ions for silicon. On the other hand, unusual conditions may result in passivation of materials that are normally unprotected, as the alkaline environment of concrete does for steel rebar. Exposure to a liquid metal such as mercury or hot solder can often circumvent passivation mechanisms.

Corrosion in Passivated Materials

Passivation is extremely useful in mitigating corrosion damage, however even a high-quality alloy will corrode if its ability to form a passivating film is hindered. Proper selection of the right grade of material for the specific environment is important for the long-lasting performance of this group of materials. If breakdown occurs in the passive film due to chemical or mechanical factors, the resulting major modes of corrosion may include pitting corrosion, crevice corrosion, and stress corrosion cracking.

Pitting Corrosion

Certain conditions, such as low concentrations of oxygen or high concentrations of species such as chloride which complete as anions, can interfere with a given alloy's ability to re-form a passivating film. In the worst case, almost all of the surface will remain protected, but tiny local fluctuations will degrade the oxide film in a few critical points. Corrosion at these points will be greatly amplified, and can cause *corrosion pits* of several types, depending upon conditions. While the corrosion pits only nucleate under fairly extreme circumstances, they can continue to grow even when conditions return to normal, since the interior of a pit is naturally deprived of oxygen and locally the pH decreases to very low values and the corrosion rate increases due to an autocatalytic process. In extreme cases, the sharp tips of extremely long and narrow corrosion pits can cause stress concentration to the point that otherwise tough alloys can shatter; a thin film pierced by an invisibly small hole can hide a thumb sized pit from view. These problems are especially dangerous because they are difficult to detect before a part or structure fails. Pitting remains among the most common and damaging forms of corrosion in passivated alloys, but it can be prevented by control of the alloy's environment.

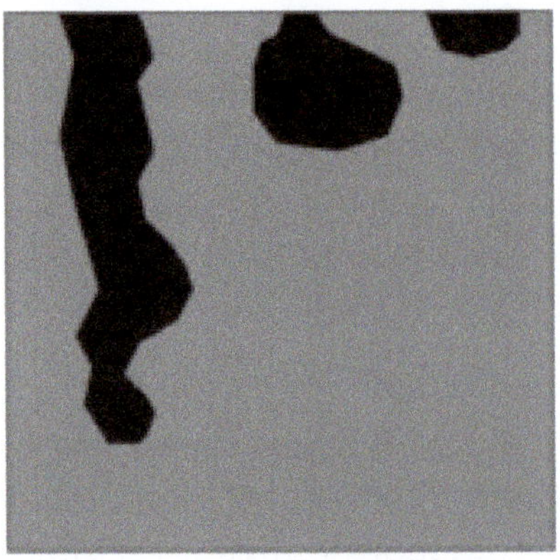

The scheme of pitting corrosion

Pitting results when a small hole, or cavity, forms in the metal, usually as a result of de-passivation of a small area. This area becomes anodic, while part of the remaining metal becomes cathodic, producing a localized galvanic reaction. The deterioration of this small area penetrates the metal and can lead to failure. This form of corrosion is often difficult to detect due to the fact that it is usually relatively small and may be covered and hidden by corrosion-produced compounds.

Weld Decay and Knifeline Attack

Stainless steel can pose special corrosion challenges, since its passivating behavior relies on the presence of a major alloying component (chromium, at least 11.5%). Because of the elevated temperatures of welding and heat treatment, chromium carbides can form in the grain boundaries of stainless alloys. This chemical reaction robs the material of chromium in the zone near the grain boundary, making those areas much less resistant to corrosion. This creates a galvanic couple with the well-protected alloy nearby, which leads to *weld decay* (corrosion of the grain boundaries in the heat affected zones) in highly corrosive environments.

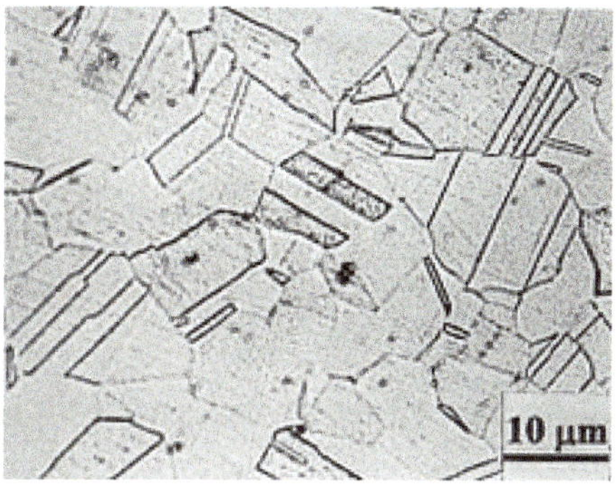

Normal microstructure

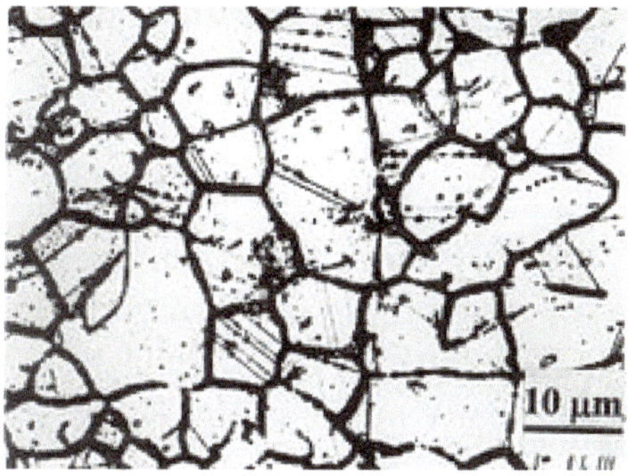

Sensitized microstructure

A stainless steel is said to be sensitized if chromium carbides are formed in the microstructure. A typical microstructure of a normalized type 304 stainless steel shows no signs of sensitization while a heavily sensitized steel shows the presence of grain boundary precipitates. The dark lines in the sensitized microstructure are networks of chromium carbides formed along the grain boundaries.

Special alloys, either with low carbon content or with added carbon "getters" such as titanium and niobium (in types 321 and 347, respectively), can prevent this effect, but the latter require special heat treatment after welding to prevent the similar phenomenon of *knifeline attack*. As its name implies, corrosion is limited to a very narrow zone adjacent to the weld, often only a few micrometers across, making it even less noticeable.

Crevice Corrosion

Crevice corrosion is a localized form of corrosion occurring in confined spaces (crevices), to which the access of the working fluid from the environment is limited. Formation of a differential aeration cell leads to corrosion inside the crevices. Examples of crevices are gaps and contact areas between parts, under gaskets or seals, inside cracks and seams, spaces filled with deposits and under sludge piles.

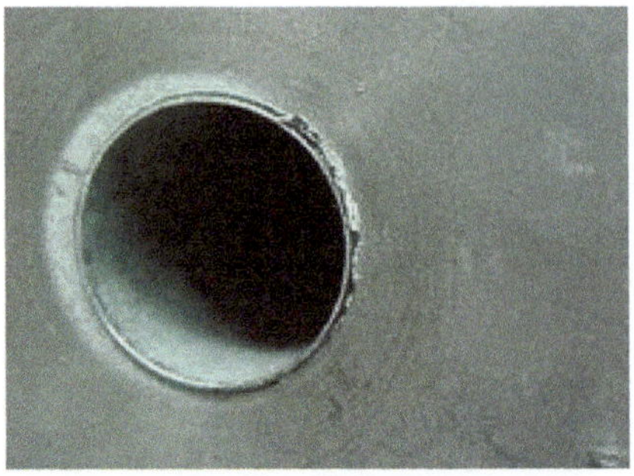

Corrosion in the crevice between the tube and tube sheet (both made of type 316 stainless steel) of a heat exchanger in a seawater desalination plant.

Crevice corrosion is influenced by the crevice type (metal-metal, metal-nonmetal), crevice geometry (size, surface finish), and metallurgical and environmental factors. The susceptibility to crevice corrosion can be evaluated with ASTM standard procedures. A critical crevice corrosion temperature is commonly used to rank a material's resistance to crevice corrosion.

Microbial Corrosion

Microbial corrosion, or commonly known as microbiologically influenced corrosion (MIC), is a corrosion caused or promoted by microorganisms, usually chemoautotrophs. It can apply to both metallic and non-metallic materials, in the presence or absence of oxygen. Sulfate-reducing bacteria are active in the absence of oxygen (anaerobic); they produce hydrogen sulfide, causing sulfide stress cracking. In the presence of oxygen (aerobic), some bacteria may directly oxidize iron to iron oxides and hydroxides, other bacteria oxidize sulfur and produce sulfuric acid causing biogenic sulfide corrosion. Concentration cells can form in the deposits of corrosion products, leading to localized corrosion.

Accelerated low-water corrosion (ALWC) is a particularly aggressive form of MIC that affects steel piles in seawater near the low water tide mark. It is characterized by an orange sludge, which smells of hydrogen sulfide when treated with acid. Corrosion rates can be very high and design corrosion allowances can soon be exceeded leading to premature failure of the steel pile. Piles that have been coated and have cathodic protection installed at the time of construction are not susceptible to ALWC. For unprotected piles, sacrificial anodes can be installed local to the affected areas to inhibit the corrosion or a complete retrofitted sacrificial anode system can be installed. Affected areas can also be treated electrochemically by using an electrode to first produce chlorine to kill the bacteria, and then to produce a calcareous deposit, which will help shield the metal from further attack.

High-temperature Corrosion

High-temperature corrosion is chemical deterioration of a material (typically a metal) as a result of heating. This non-galvanic form of corrosion can occur when a metal is subjected to a hot atmosphere containing oxygen, sulfur, or other compounds capable of oxidizing (or assisting the oxidation of) the material concerned. For example, materials used in aerospace, power generation and even in car engines have to resist sustained periods at high temperature in which they may be exposed to an atmosphere containing potentially highly corrosive products of combustion.

The products of high-temperature corrosion can potentially be turned to the advantage of the engineer. The formation of oxides on stainless steels, for example, can provide a protective layer preventing further atmospheric attack, allowing for a material to be used for sustained periods at both room and high temperatures in hostile conditions. Such high-temperature corrosion products, in the form of compacted oxide layer glazes, prevent or reduce wear during high-temperature sliding contact of metallic (or metallic and ceramic) surfaces.

Metal Dusting

Metal dusting is a catastrophic form of corrosion that occurs when susceptible materials are exposed to environments with high carbon activities, such as synthesis gas and other high-CO environments. The corrosion manifests itself as a break-up of bulk metal to metal powder. The suspected mechanism is firstly the deposition of a graphite layer on the surface of the metal, usually from carbon monoxide (CO) in the vapor phase. This graphite layer is then thought to form metastable M_3C species (where M is the metal), which migrate away from the metal surface. However, in some regimes no M_3C species is observed indicating a direct transfer of metal atoms into the graphite layer.

Protection from Corrosion

The US Army shrink wraps equipment such as helicopters to protect them from corrosion and thus save millions of dollars.

Surface Treatments

Applied Coatings

Plating, painting, and the application of enamel are the most common anti-corrosion treatments. They work by providing a barrier of corrosion-resistant material between the damaging environment and the structural material. Aside from cosmetic and manufacturing issues, there may be tradeoffs in mechanical flexibility versus resistance to abrasion and high temperature. Platings usually fail only in small sections, but if the plating is more noble than the substrate (for example, chromium on steel), a galvanic couple will cause any exposed area to corrode much more rapidly than an unplated surface would. For this reason, it is often wise to plate with active metal such as zinc or cadmium.

Galvanized surface

Painting either by roller or brush is more desirable for tight spaces; spray would be better for larger coating areas such as steel decks and waterfront applications. Flexible polyurethane coatings, like Durabak-M26 for example, can provide an anti-corrosive seal with a highly durable slip resistant membrane. Painted coatings are relatively easy to apply and have fast drying times although temperature and humidity may cause dry times to vary.

Reactive Coatings

If the environment is controlled (especially in recirculating systems), corrosion inhibitors can often be added to it. These chemicals form an electrically insulating or chemically impermeable coating on exposed metal surfaces, to suppress electrochemical reactions. Such methods make the system less sensitive to scratches or defects in the coating, since extra inhibitors can be made available wherever metal becomes exposed. Chemicals that inhibit corrosion include some of the salts in hard water (Roman water systems are famous for their mineral deposits), chromates, phosphates, polyaniline, other conducting polymers and a wide range of specially-designed chemicals that resemble surfactants (i.e. long-chain organic molecules with ionic end groups).

Anodization

Aluminium alloys often undergo a surface treatment. Electrochemical conditions in the bath are carefully adjusted so that uniform pores, several nanometers wide, appear in the metal's oxide film. These pores allow the oxide to grow much thicker than passivating conditions would allow. At the end of the treatment, the pores are allowed to seal, forming a harder-than-usual surface layer. If this coating is scratched, normal passivation processes take over to protect the damaged area.

This climbing descender is anodized with a yellow finish.

Anodizing is very resilient to weathering and corrosion, so it is commonly used for building facades and other areas where the surface will come into regular contact with the elements. While being resilient, it must be cleaned frequently. If left without cleaning, panel edge staining will naturally occur. Anodization is the process of converting an anode into cathode by bringing a more active anode in contact with it.

Biofilm Coatings

A new form of protection has been developed by applying certain species of bacterial films to the surface of metals in highly corrosive environments. This process increases the corrosion resistance substantially. Alternatively, antimicrobial-producing biofilms can be used to inhibit mild steel corrosion from sulfate-reducing bacteria.

Controlled Permeability Formwork

Controlled permeability formwork (CPF) is a method of preventing the corrosion of reinforcement by naturally enhancing the durability of the cover during concrete placement. CPF has been used in environments to combat the effects of carbonation, chlorides, frost and abrasion.

Cathodic Protection

Cathodic protection (CP) is a technique to control the corrosion of a metal surface by making that surface the cathode of an electrochemical cell. Cathodic protection systems are most commonly used to protect steel, and pipelines and tanks; steel pier piles, ships, and offshore oil platforms.

Sacrificial Anode Protection

For effective CP, the potential of the steel surface is polarized (pushed) more negative until the metal surface has a uniform potential. With a uniform potential, the driving force for the corrosion reaction is halted. For galvanic CP systems, the anode material corrodes under the influence of the steel, and eventually it must be replaced. The polarization is caused by the current flow from the anode to the cathode, driven by the difference in electrode potential between the anode and the cathode.

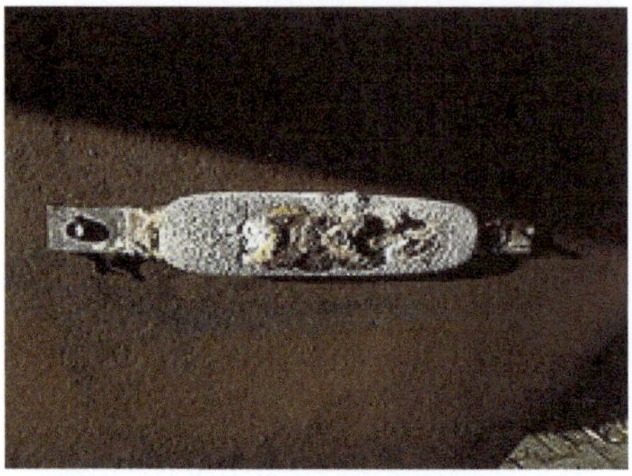

Sacrificial anode on the hull of a ship.

Impressed Current Cathodic Protection

For larger structures, galvanic anodes cannot economically deliver enough current to provide complete protection. Impressed current cathodic protection (ICCP) systems use anodes connected to a DC power source (such as a cathodic protection rectifier). Anodes for ICCP systems are tubular and solid rod shapes of various specialized materials. These include high silicon cast iron, graphite, mixed metal oxide or platinum coated titanium or niobium coated rod and wires.

Anodic Protection

Anodic protection impresses anodic current on the structure to be protected (opposite to the cathodic protection). It is appropriate for metals that exhibit passivity (e.g. stainless steel) and suitably small passive current over a wide range of potentials. It is used in aggressive environments, such as solutions of sulfuric acid.

Rate of Corrosion

A simple test for measuring corrosion is the weight loss method. The method involves exposing a clean weighed piece of the metal or alloy to the corrosive environment for a specified time followed by cleaning to remove corrosion products and weighing the piece to determine the loss of weight. The rate of corrosion (R) is calculated as

$$R = \frac{kW}{\rho A t}$$

where k is a constant, W is the weight loss of the metal in time t, A is the surface area of the metal exposed, and ρ is the density of the metal (in g/cm^3).

Other common expressions for the corrosion rate is penetration depth and change of mechanical properties.

Economic Impact

In 2002, the US Federal Highway Administration released a study titled "Corrosion Costs and Preventive Strategies in the United States" on the direct costs associated with metallic corrosion in the US industry. In 1998, the total annual direct cost of corrosion in the U.S. was ca. \$276 billion (ca. 3.2% of the US gross domestic product). Broken down into five specific industries, the economic losses are \$22.6 billion in infrastructure; \$17.6 billion in production and manufacturing; \$29.7 billion in transportation; \$20.1 billion in government; and \$47.9 billion in utilities.

The collapsed Silver Bridge, as seen from the Ohio side

Rust is one of the most common causes of bridge accidents. As rust has a much higher volume than the originating mass of iron, its build-up can also cause failure by forcing apart adjacent parts. It was the cause of the collapse of the Mianus river bridge in 1983, when the bearings rusted internally and pushed one corner of the road slab off its support. Three drivers on the roadway at the time died as the slab fell into the river below. The following NTSB investigation showed that a drain in the road had been blocked for road re-surfacing, and had not been unblocked; as a result, runoff water penetrated the support hangers. Rust was also an important factor in the Silver Bridge disaster of 1967 in West Virginia, when a steel suspension bridge collapsed within a minute, killing 46 drivers and passengers on the bridge at the time.

Similarly, corrosion of concrete-covered steel and iron can cause the concrete to spall, creating severe structural problems. It is one of the most common failure modes of reinforced concrete bridges. Measuring instruments based on the half-cell potential can detect the potential corrosion spots before total failure of the concrete structure is reached.

Until 20–30 years ago, galvanized steel pipe was used extensively in the potable water systems for single and multi-family residents as well as commercial and public construction. Today, these systems have long ago consumed the protective zinc and are corroding internally resulting in poor water quality and pipe failures. The economic impact on homeowners, condo dwellers, and the public infrastructure is estimated at 22 billion dollars as the insurance industry braces for a wave of claims due to pipe failures.

Corrosion in Nonmetals

Most ceramic materials are almost entirely immune to corrosion. The strong chemical bonds that hold them together leave very little free chemical energy in the structure; they can be thought of as already corroded. When corrosion does occur, it is almost always a simple dissolution of the material or chemical reaction, rather than an electrochemical process. A common example of corrosion protection in ceramics is the lime added to soda-lime glass to reduce its solubility in water; though it is not nearly as soluble as pure sodium silicate, normal glass does form sub-microscopic flaws when exposed to moisture. Due to its brittleness, such flaws cause a dramatic reduction in the strength of a glass object during its first few hours at room temperature.

Corrosion of Polymers

Polymer degradation involves several complex and often poorly understood physiochemical processes. These are strikingly different from the other processes discussed here, and so the term "corrosion" is only applied to them in a loose sense of the word. Because of their large molecular weight, very little entropy can be gained by mixing a given mass of polymer with another substance, making them generally quite difficult to dissolve. While dissolution is a problem in some polymer applications, it is relatively simple to design against.

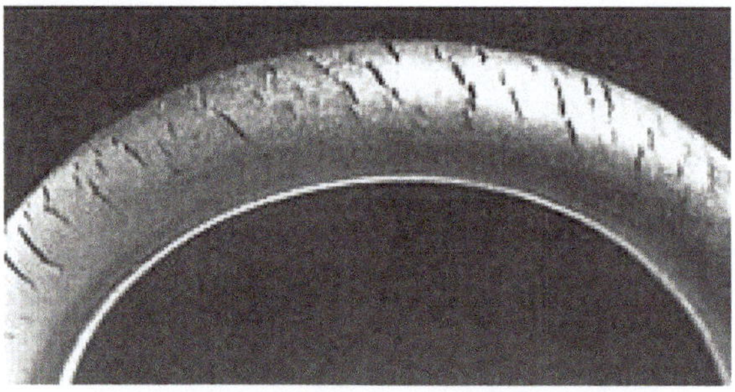

Ozone cracking in natural rubber tubing

A more common and related problem is *swelling*, where small molecules infiltrate the structure, reducing strength and stiffness and causing a volume change. Conversely, many polymers (notably flexible vinyl) are intentionally swelled with plasticizers, which can be leached out of the structure, causing brittleness or other undesirable changes.

The most common form of degradation, however, is a decrease in polymer chain length. Mechanisms which break polymer chains are familiar to biologists because of their effect on DNA: ionizing radiation (most commonly ultraviolet light), free radicals, and oxidizers such as oxygen, ozone, and chlorine. Ozone cracking is a well-known problem affecting natural rubber for example. Additives can slow these process very effectively, and can be as simple as a UV-absorbing pigment (e.g. titanium dioxide or carbon black). Plastic shopping bags often do not include these additives so that they break down more easily as ultrafine particles of litter.

Corrosion of Glasses

Glass is characterized by a high degree of corrosion-resistance. Because of its high water-resistance it is often used as primary packaging material in the pharma industry since most medicines are preserved in a watery solution. Besides its water-resistance, glass is also very robust when being exposed to chemically aggressive liquids or gases. While other materials like metal or plastics quickly reach their limits, special glass-types can easily hold up.

Glass corrosion

Glass disease is the corrosion of silicate glasses in aqueous solutions. It is governed by two mechanisms: diffusion-controlled leaching (ion exchange) and hydrolytic dissolution of the glass network. Both mechanisms strongly depend on the pH of contacting solution: the rate of ion exchange decreases with pH as $10^{-0.5pH}$ whereas the rate of hydrolytic dissolution increases with pH as $10^{0.5pH}$.

Mathematically, corrosion rates of glasses are characterized by normalized corrosion rates of elements NR_i (g/cm²·d) which are determined as the ratio of total amount of released species into the water M_i (g) to the water-contacting surface area S (cm²), time of contact t (days) and weight fraction content of the element in the glass f_i:

$$\mathrm{NR}_i = \frac{M_i}{Sf_i t}.$$

The overall corrosion rate is a sum of contributions from both mechanisms (leaching + dissolution) $NR_i = NRx_i + NRh$. Diffusion-controlled leaching (ion exchange) is characteristic of the initial phase of corrosion and involves replacement of alkali ions in the glass by a hydronium (H_3O^+) ion from the solution. It causes an ion-selective depletion of near surface layers of glasses and gives an inverse square root dependence of corrosion rate with exposure time. The diffusion-controlled normalized leaching rate of cations from glasses (g/cm²·d) is given by:

$$\mathrm{NR}x_i = 2\rho\sqrt{\frac{D_i}{\pi t}},$$

where t is time, D_i is the i-th cation effective diffusion coefficient (cm²/d), which depends on pH of contacting water as $D_i = D_{io} \cdot 10^{-pH}$, and ρ is the density of the glass (g/cm³).

Glass network dissolution is characteristic of the later phases of corrosion and causes a congruent release of ions into the water solution at a time-independent rate in dilute solutions (g/cm²·d):

$$\mathrm{NR}h = \rho r_h,$$

where r_h is the stationary hydrolysis (dissolution) rate of the glass (cm/d). In closed systems the consumption of protons from the aqueous phase increases the pH and causes a fast transition to hydrolysis. However, a further saturation of solution with silica impedes hydrolysis and causes the glass to return to an ion-exchange, e.g. diffusion-controlled regime of corrosion.

In typical natural conditions normalized corrosion rates of silicate glasses are very low and are of the order of 10^{-7}–10^{-5} g/(cm²·d). The very high durability of silicate glasses in water makes them suitable for hazardous and nuclear waste immobilisation.

Glass Corrosion Tests

There exist numerous standardized procedures for measuring the corrosion (also called chemical durability) of glasses in neutral, basic, and acidic environments, under simulated environmental conditions, in simulated body fluid, at high temperature and pressure, and under other conditions.

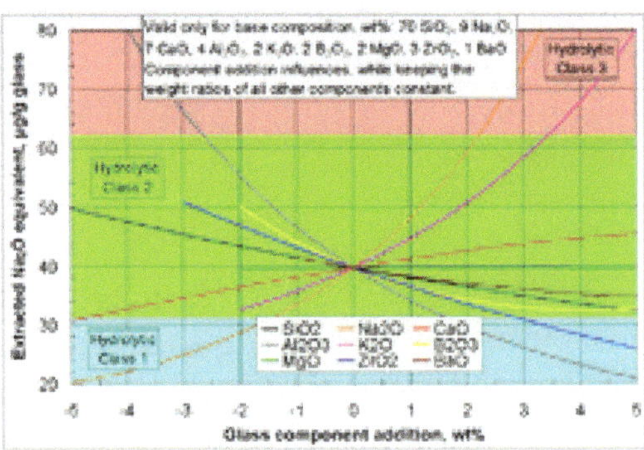

Effect of addition of a certain glass component on the chemical durability against water corrosion of a specific base glass (corrosion test ISO 719).

The standard procedure ISO 719 describes a test of the extraction of water-soluble basic compounds under neutral conditions: 2 g of glass, particle size 300–500 µm, is kept for 60 min in 50 ml de-ionized water of grade 2 at 98 °C; 25 ml of the obtained solution is titrated against 0.01 mol/l HCl solution. The volume of HCl required for neutralization is classified according to the table below.

Amount of 0.01M HCl needed to neutralize extracted basic oxides, ml	Extracted Na_2O equivalent, µg	Hydrolytic class
< 0.1	< 31	1
0.1-0.2	31-62	2
0.2-0.85	62-264	3
0.85-2.0	264-620	4
2.0-3.5	620-1085	5
> 3.5	> 1085	> 5

The standardized test ISO 719 is not suitable for glasses with poor or not extractable alkaline components, but which are still attacked by water, e.g. quartz glass, B_2O_3 glass or P_2O_5 glass.

Usual glasses are differentiated into the following classes:

Hydrolytic class 1 (Type I):

This class, which is also called neutral glass, includes borosilicate glasses (e.g. Duran, Pyrex, Fiolax).

Glass of this class contains essential quantities of boron oxides, aluminium oxides and alkaline earth oxides. Through its composition neutral glass has a high resistance against temperature shocks and the highest hydrolytic resistance. Against acid and neutral solutions it shows high chemical resistance, because of its poor alkali content against alkaline solutions.

Hydrolytic class 2 (Type II):

This class usually contains sodium silicate glasses with a high hydrolytic resistance through surface finishing. Sodium silicate glass is a silicate glass, which contains alkali- and alkaline earth oxide and primarily sodium oxide and Calcium oxide.

Hydrolytic class 3 (Type III):

Glass of the 3rd hydrolytic class usually contains sodium silicate glasses and has a mean hydrolytic resistance, which is two times poorer than of type 1 glasses.

Acid class DIN 12116 and alkali class DIN 52322 (ISO 695) are to be distinguished from the hydrolytic class DIN 12111 (ISO 719).

Pitting Corrosion

Pitting corrosion, or pitting, is a form of extremely localized corrosion that leads to the creation of small holes in the metal. The driving power for pitting corrosion is the depassivation of a small area, which becomes anodic while an unknown but potentially vast area becomes cathodic, leading to very localized galvanic corrosion. The corrosion penetrates the mass of the metal, with limited diffusion of ions. The mechanism of pitting corrosion is probably the same as crevice corrosion.

Mechanism

The more conventional explanation for pitting corrosion is that it is an autocatalytic process. Metal oxidation results in localized acidity that is maintained by the spatial separation of the cathodic and anodic half-reactions, which creates a potential gradient and electromigration of aggressive anions into the pit. Take the instance of a metal in an oxygenated NaCl electrolyte. The pit is anodic and the metal surface is cathodic. The localized production of positive metal ions in the pit gives a local excess of positive charge which attracts the negative chlorine ions from the electrolyte to produce charge neutrality. The pit contains a high concentration of MCl molecules which react with water to produce HCl, the metal hydroxide, and H+ ions, accelerating the corrosion process.In the pit, the oxygen concentration is essentially zero and all of the cathodic oxygen reactions take place on the metal surface outside the pit. The pit is anodic and the locus of rapid dissolution of the metal.

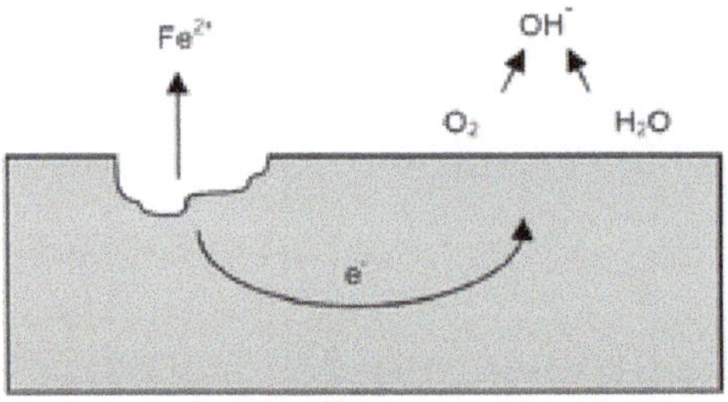

Diagram showing a mechanism of localized corrosion developing on metal in a solution containing oxygen

This kind of corrosion is extremely insidious, as it causes little loss of material with small effect on its surface, while it damages the deep structures of the metal. The pits on the surface are often obscured by corrosion products.

Pitting can be initiated by a small surface defect, being a scratch or a local change in composition, or a damage to protective coating. Polished surfaces display higher resistance to pitting.

Susceptible Alloy / Environment Combinations

Pitting Corrosion is defined by localized attack (microns - millimeters in diameter) in an otherwise passive surface and only occurs for specific alloy / environment combinations. Thus, this type of corrosion typically occurs in alloys that are protected by a tenacious (passivating) oxide film such as stainless steels, nickel alloys, aluminum alloys in environments that contain an aggressive species such as chlorides (Cl-). In contrast, alloy / environment combinations where the passive film is not very protective usually will not produce pitting corrosion. A good example of the importance of environment / alloy combinations is carbon steel. In environments where the pH is less than 10, carbon steel does not form a passivating oxide film and the addition of chloride results in uniform attack over the entire surface. However, at pH greater than 10 the oxide is very protecting and the addition of chloride results in pitting corrosion.

Besides chlorides, other anions implicated in pitting include thiosulfates ($S_2O_3^{2-}$), fluorides and iodides. Stagnant water conditions favour pitting. Thiosulfates are particularly aggressive species and are formed by partial oxidation of pyrite, or partial reduction of sulfate. Thiosulfates are a concern for corrosion in many industries: sulfide ores processing, oil wells and pipelines transporting soured oils, Kraft paper production plants, photographic industry, methionine and lysine factories.

Corrosion inhibitors, when present in sufficient amount, will provide protection against pitting. However, too low level of them can aggravate pitting by forming local anodes.

Engineering Failures Due to Pitting Corrosion

A single pit in a critical point can cause a great deal of damage. One example is the explosion in Guadalajara, Mexico on April 22, 1992, when gasoline fumes accumulated in sewers destroyed kilometers of streets. The vapors originated from a leak of gasoline through a single hole formed by corrosion between a steel gasoline pipe and a zinc-plated water pipe. Firearms can also suffer from pitting, most notably in the bore of the barrel when corrosive ammunition is used and the barrel is not cleaned soon afterward. Deformities in the bore caused by pitting can greatly reduce the firearm's accuracy. To prevent pitting in firearm bores, most modern firearms have a bore lined with chromium.

A corrosion pit on the outside wall of a pipeline at a coating defect before and after abrasive blasting.

The collapsed Silver Bridge, as seen from the Ohio side

Pitting corrosion can also help initiate stress corrosion cracking, as happened when a single eyebar on the Silver Bridge, West Virginia failed, killing 46 people on the bridge in December, 1967.

Intergranular Corrosion

Intergranular corrosion (IGC), also known as intergranular attack (IGA), is a form of corrosion where the boundaries of crystallites of the material are more susceptible to corrosion than their insides. (*Cf.* transgranular corrosion.)

Microscope view of a polished cross section of a material attacked by intergranular corrosion

This situation can happen in otherwise corrosion-resistant alloys, when the grain boundaries are depleted, known as *grain boundary depletion*, of the corrosion-inhibiting elements such as chromium by some mechanism. In nickel alloys and austenitic stainless steels, where chromium is added for corrosion resistance, the mechanism involved is precipitation of chromium carbide at the grain boundaries, resulting in the formation of chromium-depleted zones adjacent to the grain boundaries (this process is called sensitization). Around 12% chromium is minimally required to ensure passivation, a mechanism by which an ultra thin invisible film, known as passive film, forms on the surface of stainless steels. This passive film protects the metal from corrosive environments. The self-healing property of the passive film make the steel stainless. Selective leaching often involves grain boundary depletion mechanisms.

These zones also act as local galvanic couples, causing local galvanic corrosion. This condition happens when the material is heated to temperature around 700 °C for too long time, and often happens during welding or an improper heat treatment. When zones of such material form due to welding, the resulting corrosion is termed weld decay. Stainless steels can be stabilized against this behavior by addition of titanium, niobium, or tantalum, which form titanium carbide, niobium carbide and tantalum carbide preferentially to chromium carbide, by lowering the content of carbon in the steel and in case of welding also in the filler metal under 0.02%, or by heating the entire part above 1000 °C and quenching it in water, leading to dissolution of the chromium carbide in the grains and then preventing its precipitation. Another possibility is to keep the welded parts thin enough so that, upon cooling, the metal dissipates heat too quickly for chromium carbide to precipitate. The ASTM A923, ASTM A262, and other similar tests are often used to determine when stainless steels are susceptible to intergranular corrosion. The tests require etching with chemicals that reveal the presence of intermetallic particles, sometimes combined with Charpy V-Notch and other mechanical testing.

Another related kind of intergranular corrosion is termed knifeline attack (KLA). Knifeline attack impacts steels stabilized by niobium, such as 347 stainless steel. Titanium, niobium, and their carbides dissolve in steel at very high temperatures. At some cooling regimes (depending on the rate of cooling), niobium carbide does not precipitate and the steel then behaves like unstabilized steel, forming chromium carbide instead. This affects only a thin zone several millimeters wide in the very vicinity of the weld, making it difficult to spot and increasing the corrosion speed. Structures made of such steels have to be heated in a whole to about 1950 °F, when the chromium carbide dissolves and niobium carbide forms. The cooling rate after this treatment is not important, as the carbon that would otherwise pose risk of formation of chromium carbide is already sequestered as niobium carbide.

Aluminium-based alloys may be sensitive to intergranular corrosion if there are layers of materials acting as anodes between the aluminium-rich crystals. High strength aluminium alloys, especially when extruded or otherwise subjected to high degree of working, can undergo exfoliation corrosion, where the corrosion products build up between the flat, elongated grains and separate them, resulting in lifting or leafing effect and often propagating from edges of the material through its entire structure. Intergranular corrosion is a concern especially for alloys with high content of copper.

Other kinds of alloys can undergo exfoliation as well; the sensitivity of cupronickel increases together with its nickel content. A broader term for this class of corrosion is lamellar corrosion. Alloys of iron are susceptible to lamellar corrosion, as the volume of iron oxides is about seven times higher than the volume of original metal, leading to formation of internal tensile stresses tearing the material apart. Similar effect leads to formation of lamellae in stainless steels, due to the difference of thermal expansion of the oxides and the metal.

Copper-based alloys become sensitive when depletion of copper content in the grain boundaries occurs.

Anisotropic alloys, where extrusion or heavy working leads to formation of long, flat grains, are especially prone to intergranular corrosion.

Intergranular corrosion induced by environmental stresses is termed stress corrosion cracking. Inter granular corrosion can be detected by ultrasonic and eddy current methods.

Sensitization Effect

Sensitization refers to the precipitation of carbides at grain boundaries in a stainless steel or alloy, causing the steel or alloy to be susceptible to intergranular corrosion or intergranular stress corrosion cracking.

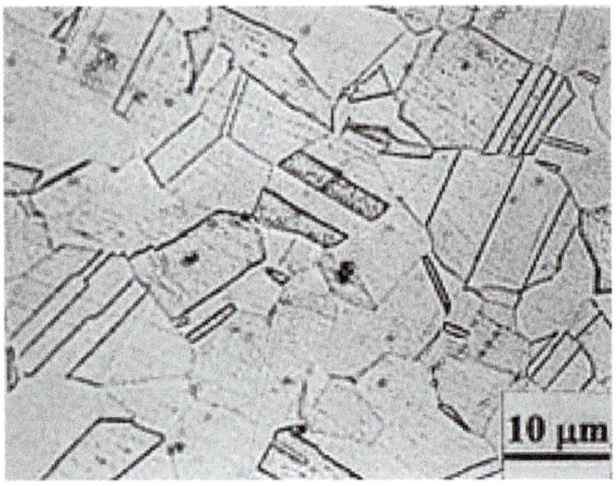

Unsensitized microstructure

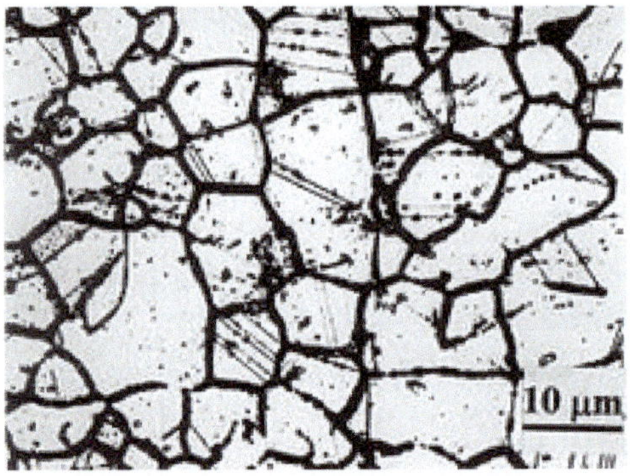

Heavily sensitized microstructure

Certain alloys when exposed to a temperature characterized as a sensitizing temperature become particularly susceptible to intergranular corrosion. In a corrosive atmosphere, the grain interfaces of these sensitized alloys become very reactive and intergranular corrosion results. This is characterized by a localized attack at and adjacent to grain boundaries with relatively little corrosion of the grains themselves. The alloy disintegrates (grains fall out) and/or loses its strength.

The photos show the typical microstructure of a normalized (unsensitized) type 304 stainless steel and a heavily sensitized steel. The samples have been polished and etched before taking the photos, and the sensitized areas show as wide, dark lines where the etching fluid has caused corrosion. The dark lines consist of carbides and corrosion products.

Intergranular corrosion is generally considered to be caused by the segregation of impurities at the grain boundaries or by enrichment or depletion of one of the alloying elements in the grain boundary areas. Thus in certain aluminium alloys, small amounts of iron have been shown to segregate in the grain boundaries and cause intergranular corrosion. Also, it has been shown that the zinc content of a brass is higher at the grain boundaries and subject to such corrosion. High-strength aluminium alloys such as the Duralumin-type alloys (Al-Cu) which depend upon precipitated phases for strengthening are susceptible to intergranular corrosion following sensitization at temperatures of about 120 °C. Nickel-rich alloys such as Inconel 600 and Incoloy 800 show similar susceptibility. Die-cast zinc alloys containing aluminum exhibit intergranular corrosion by steam in a marine atmosphere. Cr-Mn and Cr-Mn-Ni steels are also susceptible to intergranular corrosion following sensitization in the temperature range of 420°-850 °C. In the case of the austenitic stainless steels, when these steels are sensitized by being heated in the temperature range of about 520° to 800 °C, depletion of chromium in the grain boundary region occurs, resulting in susceptibility to intergranular corrosion. Such sensitization of austenitic stainless steels can readily occur because of temperature service requirements, as in steam generators, or as a result of subsequent welding of the formed structure.

Several methods have been used to control or minimize the intergranular corrosion of susceptible alloys, particularly of the austenitic stainless steels. For example, a high-temperature solution heat treatment, commonly termed solution-annealing, quench-annealing or solution-quenching, has been used. The alloy is heated to a temperature of about 1,060° to 1,120 °C and then water quenched. This method is generally unsuitable for treating large assemblies, and also ineffective where welding is subsequently used for making repairs or for attaching other structures.

Another control technique for preventing intergranular corrosion involves incorporating strong carbide formers or stabilizing elements such as niobium or titanium in the stainless steels. Such elements have a much greater affinity for carbon than does chromium; carbide formation with these elements reduces the carbon available in the alloy for formation of chromium carbides. Such a stabilized titanium-bearing austenitic chromium-nickel-copper stainless steel is shown in U.S. Pat. No. 3,562,781. Or the stainless steel may initially be reduced in carbon content below 0.03 percent so that insufficient carbon is provided for carbide formation. These techniques are expensive and only partially effective since sensitization may occur with time. The low-carbon steels also frequently exhibit lower strengths at high temperatures.

Crevice Corrosion

Crevice corrosion refers to corrosion occurring in confined spaces to which the access of the working fluid from the environment is limited. These spaces are generally called crevices. Examples of crevices are gaps and contact areas between parts, under gaskets or seals, inside cracks and seams, spaces filled with deposits and under sludge piles.

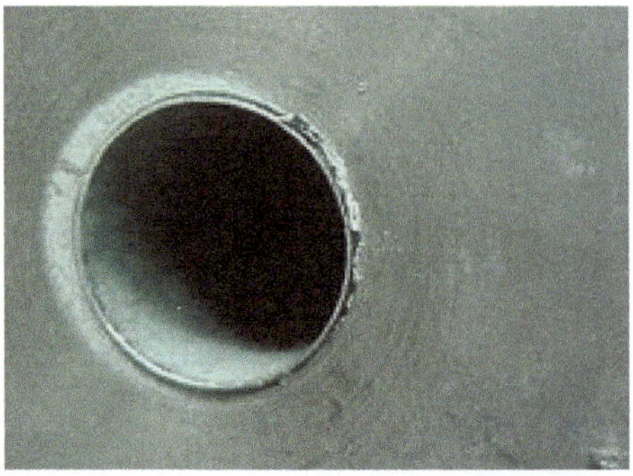

Crevice corrosion of type 316 stainless steel

This photo shows that corrosion occurred in the crevice between the tube and tube sheet (both made of type 316 stainless steel) of a heat exchanger in a sea water desalination plant.

Mechanism

The corrosion resistance of a stainless steel is dependent on the presence of an ultra-thin protective oxide film (passive film) on its surface, but it is possible under certain conditions for this oxide film to break down, for example in halide solutions or reducing acids. Areas where the oxide film can break down can also sometimes be the result of the way components are designed, for example under gaskets, in sharp re-entrant corners or associated with incomplete weld penetration or overlapping surfaces. These can all form crevices which can promote corrosion. To function as a corrosion site, a crevice has to be of sufficient width to permit entry of the corrodent, but narrow enough to ensure that the corrodent remains stagnant. Accordingly crevice corrosion usually occurs in gaps a few micrometres wide, and is not found in grooves or slots in which circulation of the corrodent is possible. This problem can often be overcome by paying attention to the design of the component, in particular to avoiding formation of crevices or at least keeping them as open as possible. Crevice corrosion is a very similar mechanism to pitting corrosion; alloys resistant to one are generally resistant to both. Crevice corrosion can be viewed as a less severe form of localized corrosion when compared with pitting. The depth of penetration and the rate of propagation in pitting corrosion are significantly greater than in crevice corrosion.

Crevices can develop a local chemistry which is very different from that of the bulk fluid. For example, in boilers, concentration of non-volatile impurities may occur in crevices near heat-transfer surfaces because of the continuous water vaporization. "Concentration factors" of many millions are not uncommon for common water impurities like sodium, sulfate or chloride. The concentration process is often referred to as "hideout" (HO), whereas the opposite process, whereby the concentrations tend to even out (e.g., during shutdown) is called "hideout return" (HOR). In a neutral pH solution, the pH inside the crevice can drop to 2, a highly acidic condition that accelerates the corrosion of most metals and alloys.

For a given crevice type, two factors are important in the initiation of crevice corrosion: the chemical composition of the electrolyte in the crevice and the potential drop into the crevice. Researchers had previously claimed that either one or the other of the two factors was responsible for initiating crevice corrosion, but recently it has been shown that it is a combination of the two that causes active crevice corrosion. Both the potential drop and the change in composition of the crevice electrolyte are caused by deoxygenation of the crevice and a separation of electroactive areas, with net anodic reactions occurring within the crevice and net cathodic reactions occurring exterior to the crevice (on the bold surface). The ratio of the surface areas between the cathodic and anodic region is significant.

Some of the phenomena occurring within the crevice may be somewhat reminiscent of galvanic corrosion:

galvanic corrosion

two connected metals + single environment

crevice corrosion

one metal part + two connected environments

The mechanism of crevice corrosion can be (but is not always) similar to that of pitting corrosion. However, there are sufficient differences to warrant a separate treatment. For example, in crevice corrosion, one has to consider the geometry of the crevice and the nature of the concentration process leading to the development of the differential local chemistry. The extreme and often unexpected local chemistry conditions inside the crevice need to be considered. Galvanic effects can play a role in crevice degradation.

Mode of Attack

Depending on the environment developed in the crevice and the nature of the metal, the crevice corrosion can take a form of:

- pitting (i.e., formation of pits),
- filiform corrosion (this type of crevice corrosion that may occur on a metallic surface underneath an organic coating),
- intergrannular attack, or
- stress corrosion cracking.

Stress Corrosion Cracking

A common form of crevice failure occurs due to stress corrosion cracking, where a crack or cracks develop from the base of the crevice where the stress concentration is greatest.

This was the root cause of the fall of the Silver Bridge in 1967 in West Virginia, where a single critical crack only about 3 mm long suddenly grew and fractured a tie bar joint. The rest of the bridge fell in less than a minute. The eyebars in the Silver Bridge were not redundant, as links were composed of only two bars each, of high strength steel (more than twice as strong as common mild steel), rather than a thick stack of thinner bars of modest material strength "combed" together as is usual for redundancy. With only two bars, the failure of one could impose excessive loading on the second, causing total failure—unlikely if more bars are used. While a low-redundancy chain can be engineered to the design requirements, the safety is completely dependent upon correct, high quality manufacturing and assembly.

The Silver Bridge upon completion in 1928

The collapsed Silver Bridge, as seen from the Ohio side

Significance

The susceptibility to crevice corrosion varies widely from one material-environment system to another. In general, crevice corrosion is of greatest concern for materials which are normally passive metals, like stainless steel or aluminum. Crevice corrosion tends to be of greatest significance to components built of highly corrosion-resistant superalloys and operating with the purest-available water chemistry. For example, steam generators in nuclear power plants degrade largely by crevice corrosion.

Crevice corrosion is extremely dangerous because it is localized and can lead to component failure while the overall material loss is minimal. The initiation and progress of crevice corrosion can be difficult to detect.

Microbial Corrosion

Microbial corrosion, also called bacterial corrosion, bio-corrosion, microbiologically influenced corrosion, or microbially induced corrosion (MIC), is corrosion caused or promoted by microorganisms, usually chemoautotrophs. It can apply to both metals and non-metallic materials.

Bacteria

Some sulfate-reducing bacteria produce hydrogen sulfide, which can cause sulfide stress cracking. *Acidithiobacillus* bacteria produce sulfuric acid; *Acidothiobacillus thiooxidans* frequently damages sewer pipes. *Ferrobacillus ferrooxidans* directly oxidizes iron to iron oxides and iron hydroxides; the rusticles forming on the RMS *Titanic* wreck are caused by bacterial activity. Other bacteria produce various acids, both organic and mineral, or ammonia.

In presence of oxygen, aerobic bacteria like *Acidithiobacillus thiooxidans*, *Thiobacillus thioparus*, and *Thiobacillus concretivorus*, all three widely present in the environment, are the common corrosion-causing factors resulting in biogenic sulfide corrosion.

Without presence of oxygen, anaerobic bacteria, especially *Desulfovibrio* and *Desulfotomaculum*, are common. *Desulfovibrio salixigens* requires at least 2.5% concentration of sodium chloride, but *D. vulgaris* and *D. desulfuricans* can grow in both fresh and salt water. *D. africanus* is another common corrosion-causing microorganism. The *Desulfotomaculum* genus comprises sulfate-reducing spore-forming bacteria; *Dtm. orientis* and *Dtm. nigrificans* are involved in corrosion processes. Sulfate-reducers require reducing environment; an electrode potential lower than -100 mV is required for them to thrive. However, even a small amount of produced hydrogen sulfide can achieve this shift, so the growth, once started, tends to accelerate.

Layers of anaerobic bacteria can exist in the inner parts of the corrosion deposits, while the outer parts are inhabited by aerobic bacteria.

Some bacteria are able to utilize hydrogen formed during cathodic corrosion processes.

Bacterial colonies and deposits can form concentration cells, causing and enhancing galvanic corrosion.

Bacterial corrosion may appear in form of pitting corrosion, for example in pipelines of the oil and gas industry. Anaerobic corrosion is evident as layers of metal sulfides and hydrogen sulfide smell. On cast iron, a graphitic corrosion selective leaching may be the result, with iron being consumed by the bacteria, leaving graphite matrix with low mechanical strength in place.

Various corrosion inhibitors can be used to combat microbial corrosion. Formulae based on benzalkonium chloride are common in oilfield industry.

Microbial corrosion can also apply to plastics, concrete, and many other materials. Two examples are Nylon-eating bacteria and Plastic-eating bacteria.

Aviation Fuel

Hydrocarbon utilizing microorganisms, mostly *Cladosporium resinae* and *Pseudomonas aeruginosa* and Sulfate Reducing Bacteria, colloquially known as "HUM bugs", are commonly present in jet fuel. They live in the water-fuel interface of the water droplets, form dark black/brown/green, gel-like mats, and cause microbial corrosion to plastic and rubber parts of the aircraft fuel system by consuming them, and to the metal parts by the means of their acidic metabolic products. They are also incorrectly called algae due to their appearance. FSII, which is added to the fuel, acts as a growth retardant for them. There are about 250 kinds of bacteria that can live in jet fuel, but fewer than a dozen are meaningfully harmful.

Nuclear Waste

Microorganisms can affect negatively radio elements confinement in nuclear waste.

Sewerage

Sewer network structures are prone to biodeterioration of materials due to the action of some microorganisms associated to the sulfur cycle. It can be a severely damaging phenomenon which was firstly described by Olmstead and Hamlin in 1900 for a brick sewer located in Los Angeles. Jointed mortar between the bricks disintegrated and ironwork was heavily rusted. The mortar joint had ballooned to two to three times its original volume, leading to the destruction or the loosening of some bricks.

Around 9% of damages described in sewer networks can be ascribed to the successive action of two kinds of microorganisms. Sulfate-reducing bacteria (SRB) can grow in relatively thick layers of sedimentary sludge and sand (typically 1 mm thick) accumulating at the bottom of the pipes and characterized by anoxic conditions. They can grow using oxidized sulfur compounds present in the effluent as electron acceptor and excrete hydrogen sulfide (H_2S). This gas is then emitted in the aerial part of the pipe and can impact the structure in two ways: either directly by reacting with the material and leading to a decrease in pH, or indirectly through its use as a nutrient by sulfur-oxidizing bacteria (SOB), growing in oxic conditions, which produce biogenic sulfuric acid. The structure is then submitted to a biogenic sulfuric acid attack. Materials like calcium aluminate cements, PVC or vitrified clay pipe may be substituted for ordinary concrete or steel sewers that are not resistant in these environments.

Anodizing

Anodizing (also spelled anodising, particularly in the UK, India and Australia) is an electrolytic passivation process used to increase the thickness of the natural oxide layer on the surface of metal parts.

These carabiners have an anodized aluminium surface that has been dyed; they are made in many colors.

The process is called *anodizing* because the part to be treated forms the anode electrode of an electrical circuit. Anodizing increases resistance to corrosion and wear, and provides better adhesion for paint primers and glues than does bare metal. Anodic films can also be used for a number of cosmetic effects, either with thick porous coatings that can absorb dyes or with thin transparent coatings that add interference effects to reflected light.

Anodizing is also used to prevent galling of threaded components and to make dielectric films for electrolytic capacitors. Anodic films are most commonly applied to protect aluminium alloys, although processes also exist for titanium, zinc, magnesium, niobium, zirconium, hafnium, and tantalum. Iron or carbon steel metal exfoliates when oxidized under neutral or alkaline microelectrolytic conditions; i.e., the iron oxide (actually ferric hydroxide or hydrated iron oxide, also known as rust) forms by anoxic anodic pits and large cathodic surface, these pits concentrate anions such as sulfate and chloride accelerating the underlying metal to corrosion. Carbon flakes or nodules in iron or steel with high carbon content (high-carbon steel, cast iron) may cause an electrolytic potential and interfere with coating or plating. Ferrous metals are commonly anodized electrolytically in nitric acid or by treatment with red fuming nitric acid to form hard black ferric oxide. This oxide remains conformal even when plated on wire and the wire is bent.

Anodizing changes the microscopic texture of the surface and the crystal structure of the metal near the surface. Thick coatings are normally porous, so a sealing process is often needed to achieve corrosion resistance. Anodized aluminium surfaces, for example, are harder than aluminium but have low to moderate wear resistance that can be improved with increasing thickness or by applying suitable sealing substances. Anodic films are generally much stronger and more adherent than most types of paint and metal plating, but also more brittle. This makes them less likely to crack and peel from aging and wear, but more susceptible to cracking from thermal stress.

History

Anodizing was first used on an industrial scale in 1923 to protect Duralumin seaplane parts from corrosion. This early chromic acid–based process was called the Bengough-Stuart process and was documented in British defence specification DEF STAN 03-24/3. It is still used today despite its legacy requirements for a complicated voltage cycle now known to be unnecessary. Variations of this process soon evolved, and the first sulfuric acid anodizing process was patented by Gower and O'Brien in 1927. Sulfuric acid soon became and remains the most common anodizing electrolyte.

Oxalic acid anodizing was first patented in Japan in 1923 and later widely used in Germany, particularly for architectural applications. Anodized aluminium extrusion was a popular architectural material in the 1960s and 1970s, but has since been displaced by cheaper plastics and powder coating. The phosphoric acid processes are the most recent major development, so far only used as pretreatments for adhesives or organic paints. A wide variety of proprietary and increasingly complex variations of all these anodizing processes continue to be developed by industry, so the growing trend in military and industrial standards is to classify by coating properties rather than by process chemistry.

Anodized Aluminium

Aluminium alloys are anodized to increase corrosion resistance and to allow dyeing (coloring), improved lubrication, or improved adhesion. However, anodizing does not increase the strength of the aluminium object. The anodic layer is non-conductive.

When exposed to air at room temperature, or any other gas containing oxygen, pure aluminium self-passivates by forming a surface layer of amorphous aluminium oxide 2 to 3 nm thick, which provides very effective protection against corrosion. Aluminium alloys typically form a thicker oxide layer, 5–15 nm thick, but tend to be more susceptible to corrosion. Aluminium alloy parts are anodized to greatly increase the thickness of this layer for corrosion resistance. The corrosion resistance of aluminium alloys is significantly decreased by certain alloying elements or impurities: copper, iron, and silicon, so 2000-, 4000-, and 6000-series Al alloys tend to be most susceptible.

Although anodizing produces a very regular and uniform coating, microscopic fissures in the coating can lead to corrosion. Further, the coating is susceptible to chemical dissolution in the presence of high- and low-pH chemistry, which results in stripping the coating and corrosion of the substrate. To combat this, various techniques have been developed either to reduce the number of fissures, to insert more chemically stable compounds into the oxide, or both. For instance, sulfuric-anodized articles are normally sealed, either through hydro-thermal sealing or precipitating sealing, to reduce porosity and interstitial pathways that allow corrosive ion exchange between the surface and the substrate. Precipitating seals enhance chemical stability but are less effective in eliminating ion exchange pathways. Most recently, new techniques to partially convert the amorphous oxide coating into more stable micro-crystalline compounds have been developed that have shown significant improvement based on shorter bond lengths.

Some aluminium aircraft parts, architectural materials, and consumer products are anodized. Anodized aluminium can be found on MP3 players, smartphones, multi-tools, flashlights, cookware, cameras, sporting goods, window frames, roofs, in electrolytic capacitors, and on many other products both for corrosion resistance and the ability to retain dye. Although anodizing only has moderate wear resistance, the deeper pores can better retain a lubricating film than a smooth surface would.

Anodized coatings have a much lower thermal conductivity and coefficient of linear expansion than aluminium. As a result, the coating will crack from thermal stress if exposed to temperatures above 80 °C (353 K). The coating can crack, but it will not peel. The melting point of aluminium oxide is 2050 °C (2323 K), much higher than pure aluminium's 658 °C (931 K). This and the non-conductivity of aluminum oxide can make welding more difficult.

In typical commercial aluminium anodizing processes, the aluminium oxide is grown down into the surface and out from the surface by equal amounts. So anodizing will increase the part dimensions on each surface by half the oxide thickness. For example, a coating that is 2 μm thick will increase the part dimensions by 1 μm per surface. If the part is anodized on all sides, then all linear dimensions will increase by the oxide thickness. Anodized aluminium surfaces are harder than aluminium but have low to moderate wear resistance, although this can be improved with thickness and sealing.

Process

The anodized aluminium layer is grown by passing a direct current through an electrolytic solution, with the aluminium object serving as the anode (the positive electrode). The current releases hydrogen at the cathode (the negative electrode) and oxygen at the surface of the aluminium anode, creating a build-up of aluminium oxide. Alternating current and pulsed current is also possible but rarely used. The voltage required by various solutions may range from 1 to 300 V DC, although most fall in the range of 15 to 21 V. Higher voltages are typically required for thicker coatings formed in sulfuric and organic acid. The anodizing current varies with the area of aluminium being anodized and typically ranges from 30 to 300 A/m^2 (2.8 to 28 A/ft^2).

Aluminium anodizing is usually performed in an acid solution, which slowly dissolves the aluminium oxide. The acid action is balanced with the oxidation rate to form a coating with nanopores, 10–150 nm in diameter. These pores are what allow the electrolyte solution and current to reach the aluminium substrate and continue growing the coating to greater thickness beyond what is produced by autopassivation. However, these same pores will later permit air or water to reach the substrate and initiate corrosion if not sealed. They are often filled with colored dyes and/or corrosion inhibitors before sealing. Because the dye is only superficial, the underlying oxide may continue to provide corrosion protection even if minor wear and scratches may break through the dyed layer.

Conditions such as electrolyte concentration, acidity, solution temperature, and current must be controlled to allow the formation of a consistent oxide layer. Harder, thicker films tend to be produced by more dilute solutions at lower temperatures with higher voltages and currents. The film thickness can range from under 0.5 micrometers for bright decorative work up to 150 micrometers for architectural applications.

Dual-finishing Aluminium

Anodizing can be performed in combination with chromate conversion coating. Each process provides corrosion resistance, with anodizing offering a significant advantage when it comes to ruggedness or physical wear resistance. The reason for combining the processes can vary, however, the significant difference between anodizing and chromate conversion coating is the electrical conductivity of the films produced. Although both stable compounds, chromate conversion coating has a greatly increased electrical conductivity. Applications where this may be useful are varied, however, the issue of grounding or earthing components as part of a larger system is an obvious one.

The dual finishing process uses the best each process has to offer, anodizing with its hard wear resistance and chromate conversion coating with its electrical conductivity.

The process steps can typically involve chromate conversion coating the entire component, followed by a masking of the surface in areas where the chromate coating must remain intact. Beyond that, the chromate coating is then dissolved in unmasked areas. The component can then be anodized, with anodizing taking to the unmasked areas. The exact process will vary dependent on service provider, component geometry and required outcome.

Other Widely Used Specifications

The most widely used anodizing specification in the US is a U.S. military spec, MIL-A-8625, which defines three types of aluminium anodization. Type I is chromic acid anodization, Type II is sulfuric acid anodization, and Type III is sulfuric acid hardcoat anodization. Other anodizing specifications include more MIL-SPECs (e.g., MIL-A-63576), aerospace industry specs by organizations such as SAE, ASTM, and ISO (e.g., AMS 2469, AMS 2470, AMS 2471, AMS 2472, AMS 2482, ASTM B580, ASTM D3933, ISO 10074, and BS 5599), and corporation-specific specs (such as those of

Boeing, Lockheed Martin, and other large contractors). AMS 2468 is obsolete. None of these specifications define a detailed process or chemistry, but rather a set of tests and quality assurance measures which the anodized product must meet. BS 1615 provides guidance in the selection of alloys for anodizing. For British defence work, a detailed chromic and sulfuric anodizing processes are described by STAN 03-24/3 and STAN 03-25/3 respectively.

Chromic Acid Anodizing (Type I)

The oldest anodizing process uses chromic acid. It is widely known as the Bengough-Stuart process. In North America it is known as Type I because it is so designated by the MIL-A-8625 standard, but it is also covered by AMS 2470 and MIL-A-8625 Type IB. In the UK it is normally specified as Def Stan 03/24 and used in areas that are prone to come into contact with propellants etc. There are also Boeing and Airbus standards. Chromic acid produces thinner, 0.5 µm to 18 µm (0.00002" to 0.0007") more opaque films that are softer, ductile, and to a degree self-healing. They are harder to dye and may be applied as a pretreatment before painting. The method of film formation is different from using sulfuric acid in that the voltage is ramped up through the process cycle.

Sulfuric Acid Anodizing (Type II & III)

Sulfuric acid is the most widely used solution to produce anodized coating. Coatings of moderate thickness 1.8 µm to 25 µm (0.00007" to 0.001") are known as Type II in North America, as named by MIL-A-8625, while coatings thicker than 25 µm (0.001") are known as Type III, hardcoat, hard anodizing, or engineered anodizing. Very thin coatings similar to those produced by chromic anodizing are known as Type IIB. Thick coatings require more process control, and are produced in a refrigerated tank near the freezing point of water with higher voltages than the thinner coatings. Hard anodizing can be made between 13 and 150 µm (0.0005" to 0.006") thick. Anodizing thickness increases wear resistance, corrosion resistance, ability to retain lubricants and PTFE coatings, and electrical and thermal insulation. Standards for thin (Soft/Standard) sulfuric anodizing are given by MIL-A-8625 Types II and IIB, AMS 2471 (undyed), and AMS 2472 (dyed), BS EN ISO 12373/1 (decorative), BS 3987 (Architectural) . Standards for thick sulfuric anodizing are given by MIL-A-8625 Type III, AMS 2469, BS 5599, BS EN 2536 and the obsolete AMS 2468 and DEF STAN 03-26/1.

Organic Acid Anodizing

Anodising can produce yellowish integral colors without dyes if it is carried out in weak acids with high voltages, high current densities, and strong refrigeration. Shades of color are restricted to a range which includes pale yellow, gold, deep bronze, brown, grey, and black. Some advanced variations can produce a white coating with 80% reflectivity. The shade of color produced is sensitive to variations in the metallurgy of the underlying alloy and cannot be reproduced consistently.

Anodising in some organic acids, for example malic acid, can enter a 'runaway' situation, in which the current drives the acid to attack the aluminum far more aggressively than normal, resulting in huge pits and scarring. Also, if the current or voltage are driven too high, 'burning' can set in; in this case the supplies act as if nearly shorted and large, uneven and amorphous black regions develop.

Integral colour anodising is generally done with organic acids, but the same effect has been produced in laboratory with very dilute sulfuric acid. Integral color anodising was originally performed with oxalic acid, but sulfonated aromatic compounds containing oxygen, particularly sulfosalicylic acid, have been more common since the 1960s. Thicknesses up to 50 μm can be achieved. Organic acid anodising is called Type IC by MIL-A-8625.

Phosphoric Acid Anodizing

Anodizing can be carried out in phosphoric acid, usually as a surface preparation for adhesives. This is described in standard ASTM D3933.

Borate and Tartrate Baths

Anodizing can also be performed in borate or tartrate baths in which aluminium oxide is insoluble. In these processes, the coating growth stops when the part is fully covered, and the thickness is linearly related to the voltage applied. These coatings are free of pores, relative to the sulfuric and chromic acid processes. This type of coating is widely used to make electrolytic capacitors, because the thin aluminium films (typically less than 0.5 μm) would risk being pierced by acidic processes.

Plasma Electrolytic Oxidation

Plasma electrolytic oxidation is a similar process, but where higher voltages are applied. This causes sparks to occur, and results in more crystalline/ceramic type coatings.

Other Metals

Anodized Titanium

Anodized titanium is used in a recent generation of dental implants. An anodized oxide layer has a thickness in the range of 30 nanometers (1.2×10^{-6} in) to several micrometers. Standards for titanium anodizing are given by AMS 2487 and AMS 2488.

Selected colors achievable through anodization of titanium.

Anodizing titanium generates an array of different colors without dyes, for which it is sometimes used in art, costume jewelry, body piercing jewelry and wedding rings. The color formed is dependent on the thickness of the oxide (which is determined by the anodizing voltage); it is caused by the interference of light reflecting off the oxide surface with light traveling through it and reflecting off the underlying metal surface.

Anodized Magnesium

Magnesium is anodized primarily as a primer for paint. A thin (5 µm) film is sufficient for this. Thicker coatings of 25 µm and up can provide mild corrosion resistance when sealed with oil, wax, or sodium silicate. Standards for magnesium anodizing are given in AMS 2466, AMS 2478, AMS 2479, and ASTM B893.

Anodized Zinc

Zinc is rarely anodized, but a process was developed by the International Lead Zinc Research Organization and covered by MIL-A-81801. A solution of ammonium phosphate, chromate and fluoride with voltages of up to 200 V can produce olive green coatings up to 80 µm thick. The coatings are hard and corrosion resistant.

Zinc or Galvanized steel can be anodized at lower voltages (20-30 V) as well using direct currents from silicate baths containing varying concentration of sodium silicate, sodium hydroxide, borax, sodium nitrite and nickel sulphate.

Anodized Niobium

Niobium anodizes in a similar fashion to titanium with a range of attractive colors being formed by interference at different film thicknesses. Again the film thickness is dependent on the anodizing voltage. Uses include jewelry and commemorative coins.

Anodized Tantalum

Tantalum anodizes in a similar fashion to titanium and niobium with a range of attractive colors being formed by interference at different film thicknesses. Again the film thickness is dependent on the anodizing voltage and typically ranges from 18-23 Angstroms per volt depending on electrolyte and temperature. Uses include Tantalum capacitors.

Dyeing

The most common anodizing processes, for example sulfuric acid on aluminium, produce a porous surface which can accept dyes easily. The number of dye colors is almost endless; however, the colors produced tend to vary according to the base alloy. Though some may prefer lighter colors, in practice they may be difficult to produce on certain alloys such as high-silicon casting grades and 2000-series aluminium-copper alloys. Another concern is the “lightfastness” of organic dyestuffs—some colors (reds and blues) are particularly prone to fading. Black dyes and gold produced by inorganic means (ferric ammonium oxalate) are more lightfast. Dyed anodizing is usually sealed to reduce or eliminate dye bleed out.

Colored iPod Mini cases are dyed following anodizing and before thermal sealing

Alternatively, metal (usually tin) can be electrolytically deposited in the pores of the anodic coating to provide colors that are more lightfast. Metal dye colors range from pale champagne to black. Bronze shades are commonly used for architectural metals.

Alternatively the color may be produced integral to the film. This is done during the anodizing process using organic acids mixed with the sulfuric electrolyte and a pulsed current.

Splash effects are created by dying the unsealed porous surface in lighter colors and then splashing darker color dyes onto the surface. Aqueous and solvent-based dye mixtures may also be alternately applied since the colored dyes will resist each other and leave spotted effects.

Printing

Photo-quality images and graphics in vivid color may be printed into the unsealed porous oxide layer using color dyes via silkscreen, sublimation transfer or digital printer. Line-art-quality graphics can be achieved by use of a printer. Color graphics may also be directly applied by hand using an airbrush, sponge or paintbrush. Printed anodizing is sealed to prevent or reduce dye bleed-out. Uses include baseball bats, signs, furniture, surgical trays, motorcycle components, and architectural moulding.

Sealing

Acidic anodizing solutions produce pores in the anodized coating. These pores can absorb dyes and retain lubricants, but are also an avenue for corrosion. When lubrication properties are not critical, they are usually sealed after dyeing to increase corrosion resistance and dye retention. Long immersion in boiling-hot deionized water or steam is the simplest sealing process, although it is not completely effective and reduces abrasion resistance by 20%. The oxide is converted into its hydrated form, and the resulting swelling reduces the porosity of the surface. Cold sealing, where the pores are closed by impregnation of a sealant in a room-temperature bath, is more popular due to energy savings. Coatings sealed in this method are not suitable for adhesive bonding. Teflon, nickel acetate, cobalt acetate, and hot sodium or potassium dichromate seals are commonly used. MIL-A-8625 requires sealing for thin coatings (Types I and II) and allows it as an option for thick ones (Type III).

Cleaning

Anodized aluminium surfaces are susceptible to panel edge staining, a unique type of surface staining that can affect the structural integrity of the metal.

Environmental Impact

Anodizing is one of the more environmentally friendly metal finishing processes. With the exception of organic (aka integral color) anodizing, the by-products contain only small amounts of heavy metals, halogens, or volatile organic compounds. Integral color anodizing produces no VOCs, heavy metals, or halogens as all of the byproducts found in the effluent streams of other processes come from their dyes or plating materials. The most common anodizing effluents, aluminium hydroxide and aluminium sulfate, are recycled for the manufacturing of alum, baking powder, cosmetics, newsprint and fertilizer or used by industrial wastewater treatment systems.

Mechanical Considerations

Anodizing will raise the surface, since the oxide created occupies more space than the base metal converted. This will generally not be of consequence except where there are tight tolerances. If so, the thickness of the anodizing layer has to be taken into account when choosing the machining dimension. A general practice on engineering drawing is to specify that "dimensions apply after all surface finishes". This will force the machine shop to take into account the anodization thickness when performing final machining of the mechanical part prior to anodization. Also in the case of small holes threaded to accept screws, anodizing may cause the screws to bind, thus the threaded holes may need to be chased with a tap to restore the original dimensions. Alternatively, special oversize taps may be used to precompensate for this growth. In the case of unthreaded holes that accept fixed-diameter pins or rods, a slightly oversized hole to allow for the dimension change may be appropriate. Depending on the alloy and thickness of the anodized coating, the same may have a significantly negative effect on fatigue life. On the contrary, it may also increase fatigue life by preventing corrosion pitting.

Cathodic Protection

Cathodic protection (CP) is a technique used to control the corrosion of a metal surface by making it the cathode of an electrochemical cell. A simple method of protection connects the metal to be protected to a more easily corroded "sacrificial metal" to act as the anode. The sacrificial metal then corrodes instead of the protected metal. For structures such as long pipelines, where passive galvanic cathodic protection is not adequate, an external DC electrical power source is used to provide sufficient current.

Aluminium sacrificial anodes *(light colored rectangular bars)* mounted on a steel jacket structure.

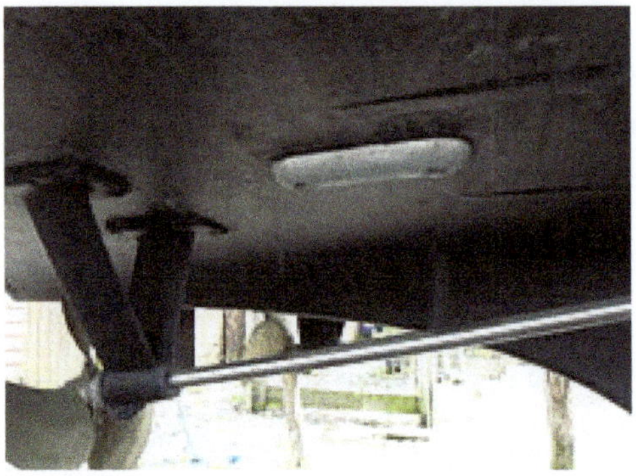

Zinc sacrificial anode *(rounded object)* screwed to the underside of the hull of a small boat.

Cathodic protection systems protect a wide range of metallic structures in various environments. Common applications are: steel water or fuel pipelines and steel storage tanks such as home water heaters; steel pier piles; ship and boat hulls; offshore oil platforms and onshore oil well casings; offshore wind farm foundations and metal reinforcement bars in concrete buildings and structures. Another common application is in galvanized steel, in which a sacrificial coating of zinc on steel parts protects them from rust.

Cathodic protection can, in some cases, prevent stress corrosion cracking.

History

Cathodic protection was first described by Sir Humphry Davy in a series of papers presented to the Royal Society in London in 1824. The first application was to the HMS

Samarang in 1824. Sacrificial anodes made from iron attached to the copper sheath of the hull below the waterline dramatically reduced the corrosion rate of the copper. However, a side effect of the cathodic protection was to increase marine growth. Copper, when corroding, releases copper ions which have an anti-fouling effect. Since excess marine growth affected the performance of the ship, the Royal Navy decided that it was better to allow the copper to corrode and have the benefit of reduced marine growth, so cathodic protection was not used further.

Davy was assisted in his experiments by his pupil Michael Faraday, who continued his research after Davy's death. In 1834, Faraday discovered the quantitative connection between corrosion weight loss and electric current and thus laid the foundation for the future application of cathodic protection.

Thomas Edison experimented with impressed current cathodic protection on ships in 1890, but was unsuccessful due to the lack of a suitable current source and anode materials. It would be 100 years after Davy's experiment before cathodic protection was used widely on oil pipelines in the United States — cathodic protection was applied to steel gas pipelines beginning in 1928 and more widely in the 1930s.

Types

Galvanic sacrificial anode attached to the hull of a ship, showing corrosion.

Galvanic

In the application of *passive* cathodic protection, a *galvanic anode*, a piece of a more electrochemically "active" metal, is attached to the vulnerable metal surface where it is exposed to an electrolyte. Galvanic anodes are selected because they have a more "active" voltage (more negative electrode potential) than the metal of the target structure (typically steel). For effective cathodic protection, the potential of the steel surface is polarized (pushed) more negative until the surface has a uniform potential. At that stage, the driving force for the corrosion reaction with the protected surface is removed. The galvanic anode continues to corrode, consuming the anode material until eventually it must be replaced. Polarization of the target structure is caused by the electron flow from the anode to the cathode, so the two metals must have a good electrically conductive contact. The driving force for the cathodic protection current is the difference in electrode potential between the anode and the cathode.

Galvanic or sacrificial anodes are made in various shapes and sizes using alloys of zinc, magnesium and aluminium. ASTM International publishes standards on the composition and manufacturing of galvanic anodes.

In order for galvanic cathodic protection to work, the anode must possess a lower (that is, more negative) electrode potential than that of the cathode (the target structure to be protected). The table below shows a simplified galvanic series which is used to select the anode metal. The anode must be chosen from a material that is lower on the list than the material to be protected.

Metal	Potential with respect to a Cu:$CuSO_4$ reference electrode in neutral pH environment (volts)
Carbon, Graphite, Coke	+0.3
Platinum	0 to -0.1
Mill scale on Steel	-0.2
High Silicon Cast Iron	-0.2
Copper, brass, bronze	-0.2
Mild steel in concrete	-0.2
Lead	-0.5
Cast iron (not graphitized)	-0.5
Mild steel (rusted)	-0.2 to -0.5
Mild steel (clean)	-0.5 to -0.8
Commercially pure aluminium	-0.8
Aluminium alloy (5% zinc)	-1.05
Zinc	-1.1
Magnesium Alloy (6% Al, 3% Zn, 0.15% Mn)	-1.6
Commercially Pure Magnesium	-1.75

Impressed Current Systems

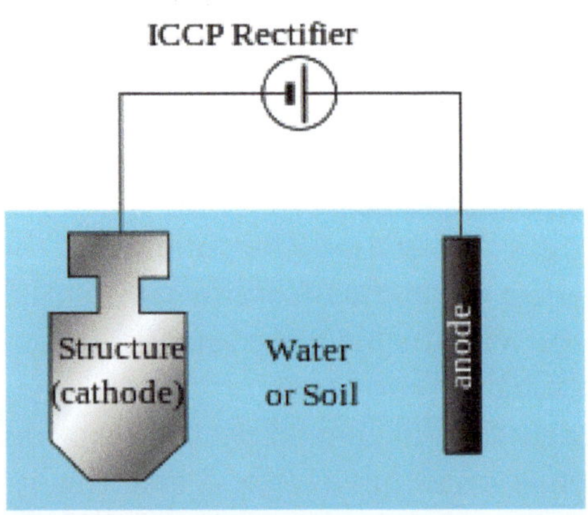

Simple impressed current cathodic protection system. A source of DC electric current is used to help drive the protective electrochemical reaction.

For larger structures, or where electrolyte resistivity is high, galvanic anodes cannot economically deliver enough current to provide protection. In these cases, *impressed current cathodic protection* (ICCP) systems are used. These consist of anodes connected to a DC power source, often a transformer-rectifier connected to AC power. In the absence of an AC supply, alternative power sources may be used, such as solar panels, wind power or gas powered thermoelectric generators.

Anodes for ICCP systems are available in a variety of shapes and sizes. Common anodes are tubular and solid rod shapes or continuous ribbons of various materials. These include high silicon cast iron, graphite, mixed metal oxide, platinum and niobium coated wire and other materials.

For pipelines, anodes are arranged in groundbeds either distributed or in a deep vertical hole depending on several design and field condition factors including current distribution requirements.

Cathodic protection transformer-rectifier units are often custom manufactured and equipped with a variety of features, including remote monitoring and control, integral current interrupters and various type of electrical enclosures. The output DC negative terminal is connected to the structure to be protected by the cathodic protection system. The rectifier output DC positive cable is connected to the anodes. The AC power cable is connected to the rectifier input terminals.

The output of the ICCP system should be optimised to provide enough current to provide protection to the target structure. Some cathodic protection transformer-rectifier units are designed with taps on the transformer windings and jumper terminals to select the voltage output of the ICCP system. Cathodic protection transformer-rectifier units for water tanks and used in other applications are made with solid state circuits to automatically adjust the operating voltage to maintain the optimum current output or structure-to-electrolyte potential. Analog or digital meters are often installed to show the operating voltage (DC and sometime AC) and current output. For shore structures and other large complex target structures, ICCP system are often designed with multiple independent zones of anodes with separate cathodic protection transformer-rectifier circuits.

Applications

Pipelines

Hazardous product pipelines are routinely protected by a coating supplemented with cathodic protection. An impressed current cathodic protection system (ICCP) for a pipeline consists of a DC power source, often an AC powered transformer rectifier and an anode, or array of anodes buried in the ground (the anode groundbed).

The DC power source would typically have a DC output of up to 50 amperes and 50 volts, but this depends on several factors, such as the size of the pipeline and coating quality. The positive DC output terminal would be connected via cables to the anode array, while another cable would connect the negative terminal of the rectifier to the pipeline, preferably through junction boxes to allow measurements to be taken.

An air cooled cathodic protection rectifier connected to a pipeline.

Anodes can be installed in a groundbed consisting of a vertical hole backfilled with conductive coke (a material that improves the performance and life of the anodes) or laid in a prepared trench, surrounded by conductive coke and backfilled. The choice of groundbed type and size depends on the application, location and soil resistivity.

The DC cathodic protection current is then adjusted to the optimum level after conducting various tests including measurements of pipe-to-soil potentials or electrode potential.

It is sometimes more economically viable to protect a pipeline using galvanic (sacrificial) anodes. This is often the case on smaller diameter pipelines of limited length. Galvanic anodes rely on the galvanic series potentials of the metals to drive cathodic protection current from the anode to the structure being protected.

Water pipelines of various pipe materials are also provided with cathodic protection where owners determine the cost is reasonable for the expected pipeline service life extension attributed to the application of cathodic protection.

Ships and Boats

The white patches visible on the ship's hull are zinc block sacrificial anodes.

Cathodic protection on ships is often implemented by galvanic anodes attached to the hull and ICCP for larger vessels. Since ships are regularly removed from the water for inspections and maintenance, it is a simple task to replace the galvanic anodes.

Galvanic anodes are generally shaped to reduced drag in the water and fitted flush to the hull to also try to minimize drag.

Smaller vessels, with non-metallic hulls, such as yachts, are equipped with galvanic anodes to protect areas such as outboard motors. As with all galvanic cathodic protection, this application relies on a solid electrical connection between the anode and the item to be protected.

For ICCP on ships, the anodes are usually constructed of a relatively inert material such as platinised titanium. A DC power supply is provided within the ship and the anodes mounted on the outside of the hull. The anode cables are introduced into the ship via a compression seal fitting and routed to the DC power source. The negative cable from the power supply is simply attached to the hull to complete the circuit. Ship ICCP anodes are flush-mounted, minimizing the effects of drag on the ship, and located a minimum 5 ft below the light load line in an area to avoid mechanical damage. The current density required for protection is a function of velocity and considered when selecting the current capacity and location of anode placement on the hull.

Some ships may require specialist treatment, for example aluminium hulls with steel fixtures will create an electrochemical cell where the aluminium hull can act as a galvanic anode and corrosion is enhanced. In cases like this, aluminium or zinc galvanic anodes can be used to offset the potential difference between the aluminium hull and the steel fixture. If the steel fixtures are large, several galvanic anodes may be required, or even a small ICCP system.

Marine

Marine cathodic protection covers many areas, jetties, harbors, offshore structures. The variety of different types of structure leads to a variety of systems to provide protection. Galvanic anodes are favored, but ICCP can also often be used. Because of the wide variety of structure geometry, composition, and architecture, specialized firms are often required to engineer structure-specific cathodic protection systems. Sometimes marine structures require retroactive modification to be effectively protected

Steel in Concrete

The application to concrete reinforcement is slightly different in that the anodes and reference electrodes are usually embedded in the concrete at the time of construction when the concrete is being poured. The usual technique for concrete buildings, bridges and similar structures is to use ICCP, but there are systems available that use the principle of galvanic cathodic protection as well, although in the UK at least, the use of galvanic anodes for atmospherically exposed reinforced concrete structures is considered experimental.

For ICCP, the principle is the same as any other ICCP system. However, in a typical atmospherically exposed concrete structure such as a bridge, there will be many more anodes distributed through the structure as opposed to an array of anodes as used on a pipeline. This makes for a more complicated system and usually an automatically controlled DC power source is used, possibly with an option for remote monitoring and operation. For buried or submerged structures, the treatment is similar to that of any other buried or submerged structure.

Galvanic systems offer the advantage of being easier to retrofit and do not need any control systems as ICCP does.

For pipelines constructed from pre-stressed concrete cylinder pipe (PCCP), the techniques used for cathodic protection are generally as for steel pipelines except that the applied potential must be limited to prevent damage to the prestressing wire.

The steel wire in a PCCP pipeline is stressed to the point that any corrosion of the wire can result in failure. An additional problem is that any excessive hydrogen ions as a result of an excessively negative potential can cause hydrogen embrittlement of the wire, also resulting in failure. The failure of too many wires will result in catastrophic failure of the PCCP. To implement ICCP therefore requires very careful control to ensure satisfactory protection. A simpler option is to use galvanic anodes, which are self-limiting and need no control.

Internal Cathodic Protection

Vessels, pipelines and tanks which are used to store or transport liquids can also be protected from corrosion on their internal surfaces by the use of cathodic protection. ICCP and galvanic systems can be used. A common application of internal cathodic protection is water storage tanks.

Galvanized Steel

Galvanizing generally refers to hot-dip galvanizing which is a way of coating steel with a layer of metallic zinc. Galvanized coatings are quite durable in most environments because they combine the barrier properties of a coating with some of the benefits of cathodic protection. If the zinc coating is scratched or otherwise locally damaged and steel is exposed, the surrounding areas of zinc coating form a galvanic cell with the exposed steel and protect it from corrosion. This is a form of localized cathodic protection - the zinc acts as a sacrificial anode.

Galvanizing, while using the electrochemical principle of cathodic protection, is not actually cathodic protection. Cathodic protection requires the anode to be separate from the metal surface to be protected, with an ionic connection through the electrolyte and an electron connection through a connecting cable, bolt or similar. This means that any area of the protected structure within the electrolyte can be protected, whereas in the case of galvanizing, only areas very close to the zinc are protected. Hence, a larger area of bare steel would only be protected around the edges.

Automobiles

Several companies market electronic corrosion control devices for automobiles and trucks and questions remain about their effectiveness without adequate scientific testing and validation. In 1996 the FTC ordered David McCready to pay restitution and banned the names "Rust Buster" and "Rust Evader."

The Auto Saver System underwent new and independent laboratory testing in 2007, commissioned by Innovation, Science and Economic Development Canada. The test was conducted by an ISO accredited laboratory under government oversight to ensure full legal and regulatory compliance, and was designed by an independent corrosion engineer from McGill University and NACE member to determine if The Auto Saver System inhibits corrosion in vehicles. The results clearly demonstrated that The Auto Saver System significantly inhibits the natural corrosion process occurring in automobiles, and have been validated by leading corrosion engineers and scientists in the United States and Canada, as well as by the Government of Canada.

Testing

Electrode potential is measured with reference electrodes. Copper-copper sulphate electrodes are used for structures in contact with soil or fresh water. Silver/silver chloride/seawater electrodes or pure zinc electrodes are used for seawater applications. The methods are described in EN 13509:2003 and NACE TM0497 along with the sources of error in the voltage that appears on the display of the meter. Interpretation of electrode potential measurements to determine the potential at the interface between the anode of the corrosion cell and the electrolyte requires training and cannot be expected to match the accuracy of measurements done in laboratory work.

Problems

Production of Hydrogen Ions

A side effect of improperly applied cathodic protection is the production of atomic hydrogen, leading to its absorption in the protected metal and subsequent hydrogen embrittlement of welds and materials with high hardness. Under normal conditions, the atomic hydrogen will combine at the metal surface to create hydrogen gas, which cannot penetrate the metal. Hydrogen atoms, however, are small enough to pass through the crystalline steel structure, and lead in some cases to hydrogen embrittlement.

Cathodic Disbonding

This is a process of disbondment of protective coatings from the protected structure (cathode) due to the formation of hydrogen ions over the surface of the protected material (cathode). Disbonding can be exacerbated by an increase in alkali ions and an increase in cathodic polarization. The degree of disbonding is also reliant on the type of coating, with some coatings affected more than others. Cathodic protection systems should be operated so that the structure does not become excessively polarized, since this also promotes disbonding due to excessively negative potentials. Cathodic disbonding occurs rapidly in pipelines that contain hot fluids because the process is accelerated by heat flow.

Cathodic Shielding

Effectiveness of cathodic protection (CP) systems on steel pipelines can be impaired by the use of solid film backed dielectric coatings such as polyethylene tapes, shrinkable pipeline sleeves, and factory applied single or multiple solid film coatings. This phenomenon occurs because of the high electrical resistivity of these film backings. Protective electric current from the cathodic protection system is blocked or shielded from reaching the underlying metal by the highly resistive film backing. Cathodic shielding was first defined in the 1980s as being a problem, and technical papers on the subject have been regularly published since then.

A 1999 report concerning a 20,600 bbl (3,280 m^3) spill from a Saskatchewan crude oil line contains an excellent definition of the cathodic shielding problem:

> "The triple situation of disbondment of the (corrosion) coating, the dielectric nature of the coating and the unique electrochemical environment established under the exterior coating, which acts as a shield to the electrical CP current, is referred to as CP shielding. The combination of tenting and disbondment permits a corrosive environment around the outside of the pipe to enter into the void between the exterior coating and the pipe surface. With the development of this CP shielding phenomenon, impressed current from the CP system cannot access exposed metal under the exterior coating to protect the pipe surface from the consequences of an aggressive corrosive environment. The CP shielding phenomenon induces changes in the potential gradient of the CP system across the exterior coating, which are further pronounced in areas of insufficient or sub-standard CP current emanating from the pipeline's CP system. This produces an area on the pipeline of insufficient CP defense against metal loss aggravated by an exterior corrosive environment."

Cathodic shielding is referenced in a number of the standards listed below. Newly issued USDOT regulation Title 49 CFR 192.112, in the section for *Additional design requirements for steel pipe using alternative maximum allowable operating pressure* requires that "The pipe must be protected against external corrosion by a non-shielding coating". Also, the NACE SP0169:2007 standard defines shielding in section 2, cautions against the use of materials that create electrical shielding in section 4.2.3, cautions against use of external coatings that create electrical shielding in section 5.1.2.3, and instructs readers to take 'appropriate action' when the effects of electrical shielding of cathodic protection current are detected on an operating pipeline in section 10.9.

Standards

- 49 CFR 192.451 - Requirements for Corrosion Control - Transportation of natural and other gas by pipeline: US minimum federal safety standards
- 49 CFR 195.551 - Requirements for Corrosion Control - Transportation of hazardous liquids by pipelines: US minimum federal safety standards
- AS 2832.4 - Australian Standard for Cathodic Protection
- ASME B31Q 0001-0191
- ASTM G 8, G 42 - Evaluating Cathodic Disbondment resistance of coatings
- DNV-RP-B401 - Cathodic Protection Design - Det Norske Veritas
- EN 12068:1999 - Cathodic protection. External organic coatings for the corrosion protection of buried or immersed steel pipelines used in conjunction with cathodic protection. Tapes and shrinkable materials
- EN 12473:2000 - General principles of cathodic protection in sea water
- EN 12474:2001 - Cathodic protection for submarine pipelines
- EN 12495:2000 - Cathodic protection for fixed steel offshore structures
- EN 12499:2003 - Internal cathodic protection of metallic structures
- EN 12696:2012 - Cathodic protection of steel in concrete
- EN 12954:2001 - Cathodic protection of buried or immersed metallic structures. General principles and application for pipelines
- EN 13173:2001 - Cathodic protection for steel offshore floating structures
- EN 13174:2001 - Cathodic protection for "Harbour Installations".
- EN 13509:2003 - Cathodic protection measurement techniques
- EN 13636:2004 - Cathodic protection of buried metallic tanks and related piping
- EN 14505:2005 - Cathodic protection of complex structures

- EN 15112:2006 - External cathodic protection of well casing
- EN 15280-2013 - Evaluation of a.c. corrosion likelihood of buried pipelines
- EN 50162:2004 - Protection against corrosion by stray current from direct current systems
- BS 7361-1:1991 - Cathodic Protection
- NACE SP0169:2013 - Control of External Corrosion on Underground or Submerged Metallic Piping Systems
- NACE TM 0497 - Measurement Techniques Related to Criteria for Cathodic Protection on Underground or Submerged Metallic Piping Systems

Corrosive Substance

A corrosive substance is one that will destroy and damage other substances with which it comes into contact. It may attack a great variety of materials, including metals and various organic compounds, but people are mostly concerned with its effects on living tissue: it causes chemical burns on contact.

The international pictogram for corrosive chemicals.

Chemical Terms

The word 'corrosive' is derived from the Latin verb *corrodere*, which means 'to gnaw', indicating how these substances seem to 'gnaw' their way through flesh or other material. Sometimes the word 'caustic' is used as a synonym but 'caustic' generally refers only to strong bases, particularly alkalis, and not to acids, oxidizers, or other non-alkaline corrosives.

A low concentration of a corrosive substance is usually an irritant. Corrosion of non-living surfaces such as metals is a distinct process. For example, a water/air electrochemical cell corrodes iron to rust. In the Globally Harmonized System, both rapid corrosion of metals and chemical corrosion of skin qualify for the "corrosive" symbol.

Corrosives are different from poisons in that corrosives are immediately dangerous to the tissues they contact, whereas poisons may have systemic toxic effects that require time to become evident. Colloquially, corrosives may be called "poisons" but the concepts are technically distinct. However, there is nothing which precludes a corrosive from being a poison; there are substances that are both corrosives and poisons.

Effects on Living Tissue

Common corrosives are either strong acids, strong bases, or concentrated solutions of certain weak acids or weak bases. They can exist as any state of matter, including liquids, solids, gases, mists or vapors.

Their action on living tissue (e.g. skin, flesh and cornea) is mainly based on acid-base reactions of amide hydrolysis, ester hydrolysis and protein denaturation. Proteins (chemically composed of amide bonds) are destroyed via amide hydrolysis while lipids (many of which have ester bonds) are decomposed by ester hydrolysis. These reactions lead to chemical burns and are the mechanism of the destruction posed by corrosives.

Some corrosives possess other chemical properties which may extend their corrosive effects on living tissue. For example, sulfuric acid (H_2SO_4) at a high concentration is also a strong dehydrating agent, capable of dehydrating carbohydrates and liberating extra heat. This results in secondary thermal burns in addition to the chemical burns and may speed up its decomposing reactions on the contact surface. Some corrosives, such as nitric acid and concentrated sulfuric acid, are strong oxidizing agents as well, which significantly contributes to the extra damage caused. Hydrofluoric acid does not necessarily cause noticeable damage upon contact, but produces tissue damage and toxicity after being painlessly absorbed. Zinc chloride solutions are capable of destroying cellulose and corroding through paper and silk since the zinc cations in the solutions specifically attack hydroxyl groups, acting as a Lewis acid. This effect is not restricted to acids; so strong a base as calcium oxide, which has a strong affinity for water (forming calcium hydroxide, itself a strong and corrosive base), also releases heat capable of contributing thermal burns as well as delivering the corrosive effects of a strong alkali to moist flesh.

In addition, some corrosive chemicals, mostly acids such as hydrochloric acid and nitric acid, are volatile and can emit corrosive mists upon contact with air. Inhalation can damage the respiratory tract.

Corrosive substances are most hazardous to eyesight. A drop of a corrosive may cause blindness within 2–10 seconds through opacification or direct destruction of the cornea.

Ingestion of corrosives can induce severe consequences, including serious damage of the gastrointestinal tract, which can lead to vomiting, severe stomach aches, and death.

Common Types of Corrosive Substances

Common corrosive chemicals are classified into:

- Acids

  - Strong acids – the most common are sulfuric acid, nitric acid and hydrochloric acid (H_2SO_4, HNO_3 and HCl, respectively).
  - Some concentrated weak acids, for example formic acid and acetic acid
  - Strong Lewis acids such as anhydrous aluminum chloride and boron trifluoride
  - Lewis acids with specific reactivity, e.g. solutions of zinc chloride
  - Extremely strong acids (superacids)
- Bases
  - Caustics or alkalis, such as sodium hydroxide, potassium hydroxide, and calcium hydroxide
  - Alkali metals in the metallic form (e.g. elemental sodium), and hydrides of alkali and alkaline earth metals, such as sodium hydride, function as strong bases and hydrate to give caustics
  - Extremely strong bases (superbases) such as alkoxides, metal amides (e.g. sodium amide) and organometallic bases such as butyllithium
  - Some concentrated weak bases, such as ammonia when anhydrous or in a concentrated solution
- Dehydrating agents such as concentrated sulfuric acid, phosphorus pentoxide, calcium oxide , anhydrous zinc chloride, also elemental alkali metals
- Strong oxidizers such as concentrated hydrogen peroxide
- Electrophilic halogens: elemental fluorine, chlorine, bromine and iodine, and electrophilic salts such as sodium hypochlorite or N-chloro compounds such as chloramine-T; halide ions are not corrosive, except for fluoride
- Organic halides and organic acid halides such as acetyl chloride and benzyl chloroformate
- Acid anhydrides
- Alkylating agents such as dimethyl sulfate
- Some organic materials such as phenol ("carbolic acid")

Personal Protective Equipment

Use of personal protective equipment, including items such as protective gloves, protective aprons, acid suits, safety goggles, a face shield, or safety shoes, is normally recommended when handling corrosive substances. Users should consult a safety data sheet for the specific recommendation for the corrosive substance of interest. The material of construction of the personal protective equipment is of critical importance as well. For example, although rubber gloves and rubber aprons may be made out of a chemically resistant elastomer such as nitrile rubber, neoprene, or butyl rubber, each of these materials has different resistance to different corrosives and they should not be substituted for each other.

Uses

Some corrosive chemicals are valued for various uses, the most common of which is in household cleaning agents. For example, most drain cleaners contain either acids or alkalis due to their capabilities of dissolving greases and proteins inside water pipes such as limescale.

In chemical uses, high chemical reactivity is often desirable, as the rates of chemical reactions depend on the activity (effective concentration) of the reactive species. For instance, catalytic sulfuric acid is used in the alkylation process in an oil refinery: the activity of carbocations, the reactive intermediate, is higher with stronger acidity, and thus the reaction proceeds faster. Once used, corrosives are most often recycled or neutralized. However, there have been environmental problems with untreated corrosive effluents or accidental discharges.

References

- JE Breakell, M Siegwart, K Foster, D Marshall, M Hodgson, R Cottis, S Lyon. Management of Accelerated Low Water Corrosion in Steel Maritime Structures, Volume 634 of CIRIA Series, 2005, ISBN 0-86017-634-7
- [Fundamentals of corrosion - Mechanisms, Causes and Preventative Methods]. Philip A. Schweitzer, Taylor and Francis Group, LLC (2010) ISBN 978-1-4200-6770-5.
- M.I. Ojovan, W.E. Lee. New Developments in Glassy Nuclear Wasteforms. Nova Science Publishers, New York (2007) ISBN 1600217834.
- Corrosion of Glass, Ceramics and Ceramic Superconductors. D.E. Clark, B.K. Zoitos (eds.), William Andrew Publishing/Noyes (1992) ISBN 081551283X.
- Edwards, Joseph (1997). Coating and Surface Treatment Systems for Metals. Finishing Publications Ltd. and ASM International. pp. 34–38. ISBN 0-904477-16-9.
- Kutz, Myer. "Protective coatings for aluminum alloys". Handbook of Environmental Degradation of Materials. Norwich, NY: William Andrew. p. 353. ISBN 978-0-8155-1749-8.
- Daniel Robles. "Potable Water Pipe Condition Assessment For a High Rise Condominium in The Pacific Northwest". GSG Group, Inc. Dan Robles, PE. Retrieved 10 December 2012.
- International Organization for Standardization, Procedure 719 (1985). Iso.org (2011-01-21). Retrieved on 2012-07-15.
- Calculation of the Chemical Durability (Hydrolytic Class) of Glasses. Glassproperties.com. Retrieved on 2012-07-15.

9

Evolution of Electrochemistry

Electrochemistry, as a branch of study, has gone through several changes, and has evolved over a number of decades. In recent times, electrochemistry has become a part of the present study, including research in batteries and fuel cells. The aspects elucidated in this chapter are of vital importance, and provide a better understanding of electrochemistry.

History of Electrochemistry

Electrochemistry, a branch of chemistry, went through several changes during its evolution from early principles related to magnets in the early 16th and 17th centuries, to complex theories involving conductivity, electric charge and mathematical methods. The term *electrochemistry* was used to describe electrical phenomena in the late 19th and 20th centuries. In recent decades, electrochemistry has become an area of current research, including research in batteries and fuel cells, preventing corrosion of metals, the use of electrochemical cells to remove refractory organics and similar contaminants in wastewater electrocoagulation and improving techniques in refining chemicals with electrolysis and electrophoresis.

Background and Dawn of Electrochemistry

The 16th century marked the beginning of scientific understanding of electricity and magnetism that culminated with the production of electric power and the industrial revolution in the late 19th century

In the 1550s, English scientist William Gilbert spent 17 years experimenting with magnetism and, to a lesser extent, electricity. For his work on magnets, Gilbert became known as "The Father of Magnetism." His book *De Magnete* quickly became the standard work throughout Europe on electrical and magnetic clear distinction between magnetism and what was then called the "amber effect" (static electricity).

German physicist Otto von Guericke beside his electrical generator while conducting experiment.

In 1663, German physicist Otto von Guericke created the first electrostatic generator, which produced static electricity by applying friction. The generator was made of a large sulfur ball inside a glass globe, mounted on a shaft. The ball was rotated by means of a crank and a static electric spark was produced when a pad was rubbed against the ball as it rotated. The globe could be removed and used as an electrical source for experiments with electricity. Von Guericke used his generator to show that like charges repelled each other.

The 18th Century and Birth of Electrochemistry

In 1709, Francis Hauksbee at the Royal Society in London discovered that by putting a small amount of mercury in the glass of Von Guericke's generator and evacuating the air from it, it would glow whenever the ball built up a charge and his hand was touching the globe. He had created the first gas-discharge lamp.

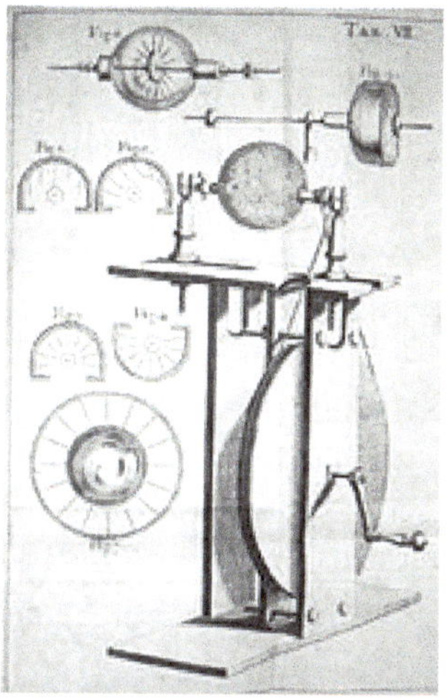

Francis Hauksbee's gas-discharge lamp

Between 1729 and 1736, two English scientists, Stephen Gray and Jean Desaguliers, performed a series of experiments which showed that a cork or other object as far away as 800 or 900 feet (245–275 m) could be electrified by connecting it via a charged glass tube to materials such as metal wires or hempen string. They found that other materials, such as silk, would not convey the effect.

By the mid-18th century, French chemist Charles François de Cisternay Du Fay had discovered two forms of static electricity, and that like charges repel each other while unlike charges attract. Du Fay announced that electricity consisted of two fluids: *vitreous* (from the Latin for "glass"), or positive, electricity; and *resinous*, or negative, electricity. This was the "two-fluid theory" of electricity, which was opposed by Benjamin Franklin's "one-fluid theory" later in the century.

In 1745, Jean-Antoine Nollet developed a theory of electrical attraction and repulsion that supposed the existence of a continuous flow of electrical matter between charged bodies. Nollet's theory at first gained wide acceptance, but met resistance in 1752 with the translation of Franklin's *Experiments and Observations on Electricity* into French. Franklin and Nollet debated the nature of electricity, with Franklin supporting action at a distance and two qualitatively opposing types of electricity, and Nollet advocating mechanical action and a single type of electrical fluid. Franklin's argument eventually won and Nollet's theory was abandoned.

In 1748, Nollet invented one of the first electrometers, the electroscope, which showed electric charge using electrostatic attraction and repulsion. Nollet is reputed to be the first to apply the name "Leyden jar" to the first device for storing electricity. Nollet's invention was replaced by Horace-Bénédict de Saussure's electrometer in 1766.

By the 1740s, William Watson had conducted several experiments to determine the speed of electricity. The general belief at the time was that electricity was faster than sound, but no accurate test been devised to measure the velocity of a current. Watson, in the fields north of London, laid out a line of wire supported by dry sticks and silk which ran for 12,276 feet (3.7 km). Even at this length, the velocity of electricity seemed instantaneous. Resistance in the wire was also noticed but apparently not fully understood, as Watson related that "we observed again, that although the electrical compositions were very severe to those who held the wires, the report of the Explosion at the prime Conductor was little, in comparison of that which is heard when the Circuit is short." Watson eventually decided not to pursue his electrical experiments, concentrating instead upon his medical career.

By the 1750s, as the study of electricity became popular, efficient ways of producing electricity were sought. The generator developed by Jesse Ramsden was among the first electrostatic generators invented. Electricity produced by such generators was used to treat paralysis, muscle spasms, and to control heart rates. Other medical uses of electricity included filling the body with electricity, drawing sparks from the body, and applying sparks from the generator to the body.

Charles-Augustin de Coulomb developed the law of electrostatic attraction in 1781 as an outgrowth of his attempt to investigate the law of electrical repulsions as stated by Joseph Priestley in England. To this end, he invented a sensitive apparatus to measure the electrical forces involved in Priestley's law. He also established the inverse square law of attraction and repulsion magnetic poles, which became the basis for the mathematical theory of magnetic forces developed by Siméon Denis Poisson. Coulomb wrote seven important works on electricity and magnetism which he submitted to the Académie des Sciences between 1785 and 1791, in which he reported having developed a theory of attraction and repulsion between charged bodies, and went on to search for perfect conductors and dielectrics. He suggested that there was no perfect dielectric, proposing that every substance has a limit, above which it will conduct electricity. The SI unit of charge is called a coulomb in his honour.

In 1789, Franz Aepinus developed a device with the properties of a "condenser" (now known as a capacitor.) The Aepinus condenser was the first capacitor developed after the Leyden jar, and was used to demonstrate conduction and induction. The device was constructed so that the space between two plates could be adjusted, and the glass dielectric separating the two plates could be removed or replaced with other materials.

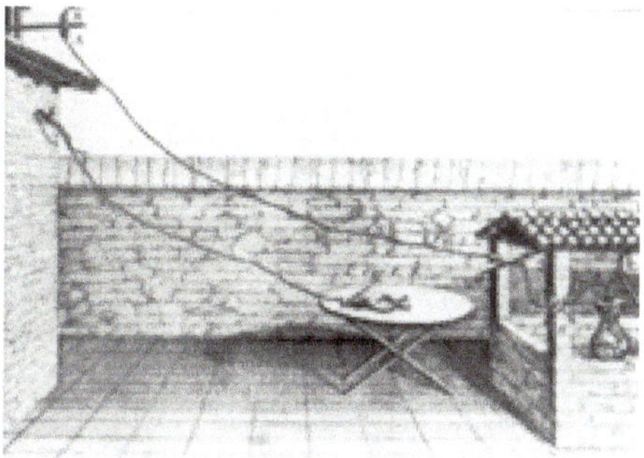

Late 1780s diagram of Galvani's experiment on frog legs.

Despite the gain in knowledge of electrical properties and the building of generators, it wasn't until the late 18th century that Italian physician and anatomist Luigi Galvani marked the birth of electrochemistry by establishing a bridge between muscular contractions and electricity with his 1791 essay *De Viribus Electricitatis in Motu Musculari Commentarius* (Commentary on the Effect of Electricity on Muscular Motion), where he proposed a "nerveo-electrical substance" in life forms.

In his essay, Galvani concluded that animal tissue contained a before-unknown innate, vital force, which he termed "animal electricity," which activated muscle when placed between two metal probes. He believed that this was evidence of a new form of electricity, separate from the "natural" form that is produced by lightning and the "artificial" form that is produced by friction (static electricity). He considered the brain to be the most important organ for the secretion of this "electric fluid" and that the nerves conducted the fluid to the muscles. He believed the tissues acted similarly to the outer and inner surfaces of Leyden jars. The flow of this electric fluid provided a stimulus to the muscle fibres.

Italian physicist Alessandro Volta showing his *"battery"* to French emperor Napoleon Bonaparte in the early 19th century.

Galvani's scientific colleagues generally accepted his views, but Alessandro Volta, the outstanding professor of physics at the University of Pavia, was not convinced by the analogy between muscles and Leyden jars. Deciding that the frogs' legs used in Galvani's experiments served only as an electroscope, he held that the contact of dissimilar metals was the true source of stimulation. He referred to the electricity so generated as "metallic electricity" and decided that the muscle, by contracting when touched by metal, resembled the action of an electroscope. Furthermore, Volta claimed that if two dissimilar metals in contact with each other also touched a muscle, agitation would also occur and increase with the dissimilarity of the metals. Galvani refuted this by obtaining muscular action using two pieces of similar metal. Volta's name was later used for the unit of electrical potential, the volt.

Rise of Electrochemistry as Branch of Chemistry

In 1800, English chemists William Nicholson and Johann Wilhelm Ritter succeeded in separating water into hydrogen and oxygen by electrolysis. Soon thereafter, Ritter discovered the process of electroplating. He also observed that the amount of metal deposited and the amount of oxygen produced during an electrolytic process depended on the distance between the electrodes. By 1801 Ritter had observed thermoelectric currents, which anticipated the discovery of thermoelectricity by Thomas Johann Seebeck.

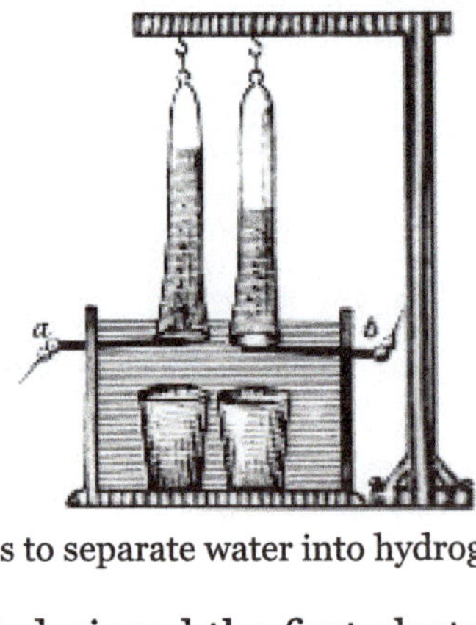

Scheme of Ritter's apparatus to separate water into hydrogen and oxygen by electrolysis

In 1802, William Cruickshank designed the first electric battery capable of mass production. Like Volta, Cruickshank arranged square copper plates, which he soldered at their ends, together with plates of zinc of equal size. These plates were placed into a long rectangular wooden box which was sealed with cement. Grooves inside the box held the metal plates in position. The box was then filled with an electrolyte of brine, or watered down acid. This flooded design had the advantage of not drying out with use and provided more energy than Volta's arrangement, which used brine-soaked papers between the plates.

In the quest for a better production of platinum metals, two scientists, William Hyde Wollaston and Smithson Tennant, worked together to design an efficient electrochemical technique to refine or purify platinum. Tennant ended up discovering the elements iridium and osmium. Wollaston's effort, in turn, led him to the discovery of the metals palladium in 1803 and rhodium in 1804.

Wollaston made improvements to the galvanic battery (named after Galvani) in the 1810s. In Wollaston's battery, the wooden box was replaced with an earthenware vessel, and a copper plate was bent into a U-shape, with a single plate of zinc placed in the center of the bent copper. The zinc plate was prevented from making contact with the copper by dowels (pieces) of cork or wood. In his single cell design, the U-shaped copper plate was welded to a horizontal handle for lifting the copper and zinc plates out of the electrolyte when the battery was not in use.

In 1809, Samuel Thomas von Soemmering developed the first telegraph. He used a device with 26 wires (1 wire for each letter of the German alphabet) terminating in a container of acid. At the sending station, a key, which completed a circuit with a battery, was connected as required to each of the line wires. The passage of current caused the acid to decompose chemically, and the message was read by observing at which of the terminals the bubbles of gas appeared. This is how he was able to send messages, one letter at a time.

Humphry Davy's work with electrolysis led to conclusion that the production of electricity in simple electrolytic cells resulted from chemical reactions between the electrolyte and the metals, and occurred between substances of opposite charge. He reasoned that the interactions of electric currents with chemicals offered the most likely means of decomposing all substances to their basic elements. These views were explained in 1806 in his lecture *On Some Chemical Agencies of Electricity*, for which he received the Napoleon Prize from the Institut de France in 1807 (despite the fact that England and France were at war at the time). This work led directly to the isolation of sodium and potassium from their common compounds and of the alkaline earth metals from theirs in 1808.

Hans Christian Ørsted's discovery of the magnetic effect of electric currents in 1820 was immediately recognised as an important advance, although he left further work on electromagnetism to others. André-Marie Ampère quickly repeated Ørsted's experiment, and formulated them mathematically (which became Ampère's law) . Ørsted also discovered that not only is a magnetic needle deflected by the electric current, but that the live electric wire is also deflected in a magnetic field, thus laying the foundation for the construction of an electric motor. Ørsted's discovery of piperine, one of the pungent components of pepper, was an important contribution to chemistry, as was his preparation of aluminium in 1825.

During the 1820s, Robert Hare developed the Deflagrator, a form of voltaic battery having large plates used for producing rapid and powerful combustion. A modified form of this apparatus was employed in 1823 in volatilising and fusing carbon. It was with these batteries that the first use of voltaic electricity for blasting under water was made in 1831.

In 1821, the Estonian-German physicist, Thomas Johann Seebeck, demonstrated the electrical potential in the juncture points of two dissimilar metals when there is a temperature difference between the joints. He joined a copper wire with a bismuth wire to form a loop or circuit. Two junctions were formed by connecting the ends of the wires to each other. He then accidentally discovered that if he heated one junction to a high temperature, and the other junction remained at room temperature, a magnetic field was observed around the circuit.

He did not recognise that an electric current was being generated when heat was applied to a bi-metal junction. He used the term "thermomagnetic currents" or "thermomagnetism" to express his discovery. Over the following two years, he reported on his continuing observations to the Prussian Academy of Sciences, where he described his observation as "the magnetic polarization of metals and ores produced by a temperature difference." This Seebeck effect became the basis of the thermocouple, which is still considered the most accurate measurement of temperature today. The converse Peltier effect was seen over a decade later when a current was run through a circuit with two dissimilar metals, resulting in a temperature difference between the metals.

In 1827 German scientist Georg Ohm expressed his law in his famous book *Die galvanische Kette, mathematisch bearbeitet* (The Galvanic Circuit Investigated Mathematically) in which he gave his complete theory of electricity.

In 1829 Antoine-César Becquerel developed the "constant current" cell, forerunner of the well-known Daniell cell. When this acid-alkali cell was monitored by a galvanometer, current was found to be constant for an hour, the first instance of "constant current". He applied the results of his study of thermoelectricity to the construction of an electric thermometer, and measured the temperatures of the interior of animals, of the soil at different depths, and of the atmosphere at different heights. He helped validate Faraday's laws and conducted extensive investigations on the electroplating of metals with applications for metal finishing and metallurgy. Solar cell technology dates to 1839 when Becquerel observed that shining light on an electrode submerged in a conductive solution would create an electric current.

Michael Faraday began, in 1832, what promised to be a rather tedious attempt to prove that all electricities had precisely the same properties and caused precisely the same effects. The key effect was electrochemical decomposition. Voltaic and electromagnetic electricity posed no problems, but static electricity did. As Faraday delved deeper into the problem, he made two startling discoveries. First, electrical force did not, as had long been supposed, act at a distance upon molecules to cause them to dissociate. It was the passage of electricity through a conducting liquid medium that caused the molecules to dissociate, even when the electricity merely discharged into the air and did not pass through a "pole" or "center of action" in a voltaic cell. Second, the amount of the decomposition was found to be related directly to the amount of electricity passing through the solution.

These findings led Faraday to a new theory of electrochemistry. The electric force, he argued, threw the molecules of a solution into a state of tension. When the force was strong enough to distort the forces that held the molecules together so as to permit the interaction with neighbouring particles, the tension was relieved by the migration of particles along the lines of tension, the different parts of atoms migrating in opposite directions. The amount of electricity that passed, then, was clearly related to the chemical affinities of the substances in solution. These experiments led directly to Faraday's two laws of electrochemistry which state:

- The amount of a substance deposited on each electrode of an electrolytic cell is directly proportional to the amount of electricity passing through the cell.
- The quantities of different elements deposited by a given amount of electricity are in the ratio of their chemical equivalent weights.

William Sturgeon built an electric motor in 1832 and invented the commutator, a ring of metal-bristled brushes which allow the spinning armature to maintain contact with the electric current and changed the alternating current to a pulsating direct current. He also improved the voltaic battery and worked on the theory of thermoelectricity.

Hippolyte Pixii, a French instrument maker, constructed the first dynamo in 1832 and later built a direct current dynamo using the commutator. This was the first practical mechanical generator of electric current that used concepts demonstrated by Faraday.

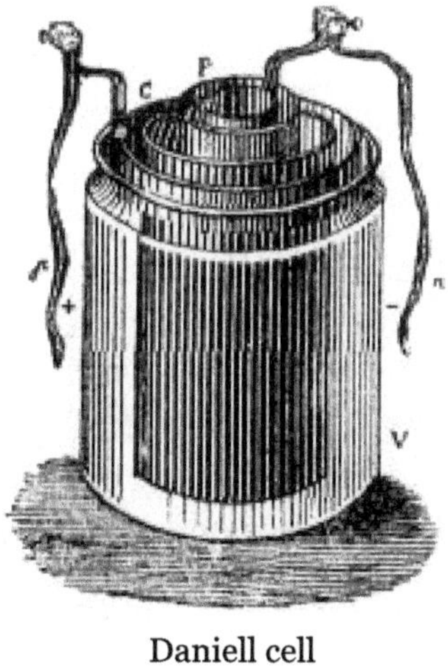

Daniell cell

John Daniell began experiments in 1835 in an attempt to improve the voltaic battery with its problems of being unsteady and a weak source of electric current. His experiments soon led to remarkable results. In 1836, he invented a primary cell in which hydrogen was eliminated in the generation of the electricity. Daniell had solved the problem of polarization. In his laboratory he had learned to alloy the amalgamated zinc of Sturgeon with mercury. His version was the first of the two-fluid class battery and the first battery that produced a constant reliable source of electric current over a long period of time.

William Grove produced the first fuel cell in 1839. He based his experiment on the fact that sending an electric current through water splits the water into its component parts of hydrogen and oxygen. So, Grove tried reversing the reaction—combining hydrogen and oxygen to produce electricity and water. Eventually the term *fuel cell* was coined in 1889 by Ludwig Mond and Charles Langer, who attempted to build the first practical device using air and industrial coal gas. He also introduced a powerful battery at the annual meeting of the British Association for the Advancement of Science in 1839. Grove's first cell consisted of zinc in diluted sulfuric acid and platinum in concentrated nitric acid, separated by a porous pot. The cell was able to generate about 12 amperes of current at about 1.8 volts. This cell had nearly double the voltage of the first Daniell cell. Grove's nitric acid cell was the favourite battery of the early American telegraph (1840–1860), because it offered strong current output.

As telegraph traffic increased, it was found that the Grove cell discharged poisonous nitrogen dioxide gas. As telegraphs became more complex, the need for a constant voltage became critical and the Grove device was limited (as the cell discharged, nitric acid was depleted and voltage was reduced). By the time of the American Civil War, Grove's battery had been replaced by the Daniell battery. In 1841 Robert Bunsen replaced the expensive platinum electrode used in Grove's battery with a carbon electrode. This led to large scale use of the "Bunsen battery" in the production of arc-lighting and in electroplating.

Wilhelm Weber developed, in 1846, the electrodynamometer, in which a current causes a coil suspended within another coil to turn when a current is passed through both. In 1852, Weber defined the absolute unit of electrical resistance (which was named the ohm after Georg Ohm). Weber's name is now used as a unit name to describe magnetic flux, the weber.

German physicist Johann Hittorf concluded that *ion movement* caused electric current. In 1853 Hittorf noticed that some ions traveled more rapidly than others. This observation led to the concept of transport number, the rate at which particular ions carried the electric current. Hittorf measured the changes in the concentration of electrolysed solutions, computed from these the transport numbers (relative carrying capacities) of many ions, and, in 1869, published his findings governing the migration of ions.

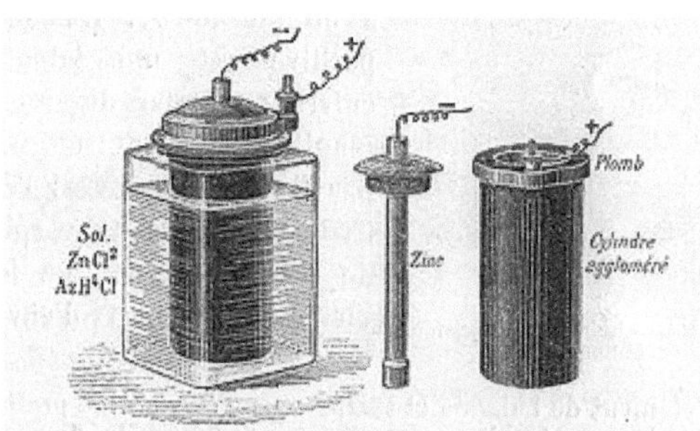

Fig. 293. — Élément Leclanché-Barbier.

Leclanché cell

In 1866, Georges Leclanché patented a new battery system, which was immediately successful. Leclanché's original cell was assembled in a porous pot. The positive electrode (the cathode) consisted of crushed manganese dioxide with a little carbon mixed in. The negative pole (anode) was a zinc rod. The cathode was packed into the pot, and a carbon rod was inserted to act as a current collector. The anode and the pot were then immersed in an ammonium chloride solution. The liquid acted as the electrolyte, readily seeping through the porous pot and making contact with the cathode material. Leclanché's "wet" cell became the forerunner to the world's first widely used battery, the zinc-carbon cell.

Late 19th Century Advances and the Advent of Electrochemical Societies

In 1869 Zénobe Gramme devised his first clean direct current dynamo. His generator featured a ring armature wound with many individual coils of wire.

Svante August Arrhenius published his thesis in 1884, *Recherches sur la conductibilité galvanique des électrolytes* (Investigations on the galvanic conductivity of electrolytes). From the results of his experiments, the author concluded that electrolytes, when dissolved in water, become to varying degrees split or dissociated into positive and negative ions. The degree to which this dissociation occurred depended above all on the nature of the substance and its concentration in the solution, being more developed the greater the dilution. The ions were supposed to be the carriers of not only the electric current, as in electrolysis, but also of the chemical activity. The relation between the actual number of ions and their number at great dilution (when all the molecules were dissociated) gave a quantity of special interest ("activity constant").

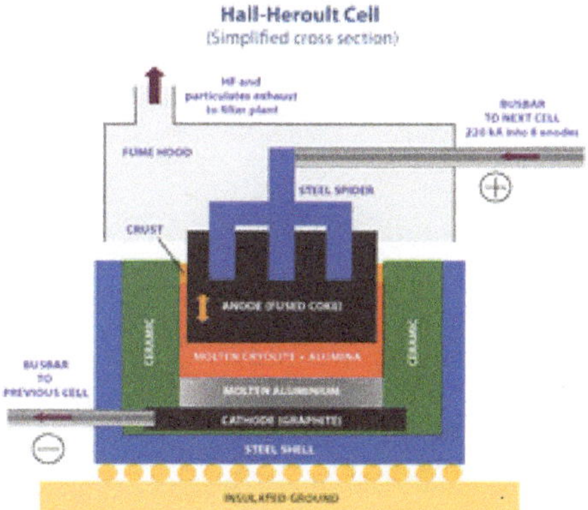

A Hall-Héroult industrial cell.

The race for the commercially viable production of aluminium was won in 1886 by Paul Héroult and Charles M. Hall. The problem many researchers had with extracting aluminium was that electrolysis of an aluminium salt dissolved in water yields aluminium hydroxide. Both Hall and Héroult avoided this problem by dissolving aluminium oxide in a new solvent— fused cryolite (Na_3AlF_6).

Wilhelm Ostwald, 1909 Nobel Laureate, started his experimental work in 1875, with an investigation on the law of mass action of water in relation to the problems of chemical affinity, with special emphasis on electrochemistry and chemical dynamics. In 1894 he gave the first modern definition of a catalyst and turned his attention to catalytic reactions. Ostwald is especially known for his contributions to the field of electrochemistry, including important studies of the electrical conductivity and electrolytic dissociation of organic acids.

Hermann Nernst developed the theory of the electromotive force of the voltaic cell in 1888. He developed methods for measuring dielectric constants and was the first to show that solvents of high dielectric constants promote the ionization of substances. Nernst's early studies in electrochemistry were inspired by Arrhenius' dissociation theory which first recognised the importance of ions in solution. In 1889, Nernst elucidated the theory of galvanic cells by assuming an "electrolytic pressure of dissolution," which forces ions from electrodes into solution and which was opposed to the osmotic pressure of the dissolved ions. He applied the principles of thermodynamics to the chemical reactions proceeding in a battery. In that same year he showed how the characteristics of the current produced could be used to calculate the free energy change in the chemical reaction producing the current. He constructed an equation, known as Nernst Equation, which describes the relation of a battery cell's voltage to its properties.

In 1898 Fritz Haber published his textbook, *Electrochemistry: Grundriss der technischen Elektrochemie auf theoretischer Grundlage* (The Theoretical Basis of Technical Electrochemistry), which was based on the lectures he gave at Karlsruhe. In the preface to his book he expressed his intention to relate chemical research to industrial processes and in the same year he reported the results of his work on electrolytic oxidation and reduction, in which he showed that definite reduction products can result if the voltage at the cathode is kept constant. In 1898 he explained the reduction of nitrobenzene in stages at the cathode and this became the model for other similar reduction processes.

In 1909, Robert Andrews Millikan began a series of experiments to determine the electric charge carried by a single electron. He began by measuring the course of charged water droplets in an electrical field. The results suggested that the charge on the droplets is a multiple of the elementary electric charge, but the experiment was not accurate enough to be convincing. He obtained more precise results in 1910 with his famous oil-drop experiment in which he replaced water (which tended to evaporate too quickly) with oil.

Jaroslav Heyrovský, a Nobel laureate, eliminated the tedious weighing required by previous analytical techniques, which used the differential precipitation of mercury by measuring drop-time. In the previous method, a voltage was applied to a dropping mercury electrode and a reference electrode was immersed in a test solution. After 50 drops of mercury were collected, they were dried and weighed. The applied voltage was varied and the experiment repeated. Measured weight was plotted versus applied voltage to obtain the curve. In 1921, Heyrovský had the idea of measuring the current flowing through the cell instead of just studying drop-time.

Heyrovský's Polarograph

On February 10, 1922, the "polarograph" was born as Heyrovský recorded the current-voltage curve for a solution of 1 mol/L NaOH. Heyrovský correctly interpreted the current increase between −1.9 and −2.0 V as being due to the deposit of Na^+ ions, forming an amalgam. Shortly thereafter, with his Japanese colleague Masuzo Shikata, he constructed the first instrument for the automatic recording of polarographic curves, which became world famous later as the polarograph.

In 1923, Johannes Nicolaus Brønsted and Thomas Martin Lowry published essentially the same theory about how acids and bases behave using electrochemical basis.

The International Society of Electrochemistry (ISE) was founded in 1949, and some years later the first sophisticated electrophoretic apparatus was developed in 1937 by Arne Tiselius, who was awarded the 1948 Nobel prize for his work in protein electrophoresis. He developed the "moving boundary," which later would become known as *zone electrophoresis*, and used it to separate serum proteins in solution. Electrophoresis became widely developed in the 1940s and 1950s when the technique was applied to molecules ranging from the largest proteins to amino acids and even inorganic ions.

During the 1960s and 1970s quantum electrochemistry was developed by Revaz Dogonadze and his pupils.

Permissions

All chapters in this book are published with permission under the Creative Commons Attribution Share Alike License or equivalent. Every chapter published in this book has been scrutinized by our experts. Their significance has been extensively debated. The topics covered herein carry significant information for a comprehensive understanding. They may even be implemented as practical applications or may be referred to as a beginning point for further studies.

We would like to thank the editorial team for lending their expertise to make the book truly unique. They have played a crucial role in the development of this book. Without their invaluable contributions this book wouldn't have been possible. They have made vital efforts to compile up to date information on the varied aspects of this subject to make this book a valuable addition to the collection of many professionals and students.

This book was conceptualized with the vision of imparting up-to-date and integrated information in this field. To ensure the same, a matchless editorial board was set up. Every individual on the board went through rigorous rounds of assessment to prove their worth. After which they invested a large part of their time researching and compiling the most relevant data for our readers.

The editorial board has been involved in producing this book since its inception. They have spent rigorous hours researching and exploring the diverse topics which have resulted in the successful publishing of this book. They have passed on their knowledge of decades through this book. To expedite this challenging task, the publisher supported the team at every step. A small team of assistant editors was also appointed to further simplify the editing procedure and attain best results for the readers.

Apart from the editorial board, the designing team has also invested a significant amount of their time in understanding the subject and creating the most relevant covers. They scrutinized every image to scout for the most suitable representation of the subject and create an appropriate cover for the book.

The publishing team has been an ardent support to the editorial, designing and production team. Their endless efforts to recruit the best for this project, has resulted in the accomplishment of this book. They are a veteran in the field of academics and their pool of knowledge is as vast as their experience in printing. Their expertise and guidance has proved useful at every step. Their uncompromising quality standards have made this book an exceptional effort. Their encouragement from time to time has been an inspiration for everyone.

The publisher and the editorial board hope that this book will prove to be a valuable piece of knowledge for students, practitioners and scholars across the globe.

Index